The Open University

Mathematics Foundation Course Unit 22

LINEAR ALGEBRA I

Prepared by the Mathematics Foundation Course Team

Correspondence Text 22

The Open University Press

Open University courses provide a method of study for independent learners through an integrated teaching system including textual material, radio and television programmes and short residential courses. This text is one of a series that make up the correspondence element of the Mathematics Foundation Course.

The Open University's courses represent a new system of university level education. Much of the teaching material is still in a developmental stage. Courses and course materials are, therefore, kept continually under revision. It is intended to issue regular up-dating notes as and when the need arises, and new editions will be brought out when necessary.

Further information on Open University courses may be obtained from The Admissions Office, The Open University, P.O. Box 48, Bletchley, Buckinghamshire.

The Open University Press
Walton Hall, Bletchley, Bucks

First Published 1971
Copyright © 1971 The Open University

Printed in Great Britain by
J W Arrowsmith Ltd, Bristol 3

SBN 335 01021 0

Contents

Objectives

The principal objective of this unit is to introduce the concept of a vector space and some related concepts.

After working through this unit you should be able to:

(i) add geometric vectors graphically and represent a scalar multiple of a geometric vector graphically;

(ii) express a given geometric vector as the sum of scalar multiples of other given geometric vectors;

(iii) explain what is meant by, and be able to calculate, the inner product of two geometric vectors;

(iv) understand what is meant by a vector space;

(v) define linear dependence and independence of a set of vectors of a vector space, and decide whether a given set of vectors is linearly dependent or independent;

(vi) define a basis of a vector space, and find bases in simple cases.

Note

Before working through this correspondence text, make sure you have read the general introduction to the mathematics course in the Study Guide, as this explains the philosophy underlying the whole course. You should also be familiar with the section which explains how a text is constructed and the meanings attached to the stars and other symbols in the margin, as this will help you to find your way through the text.

Structural Diagram

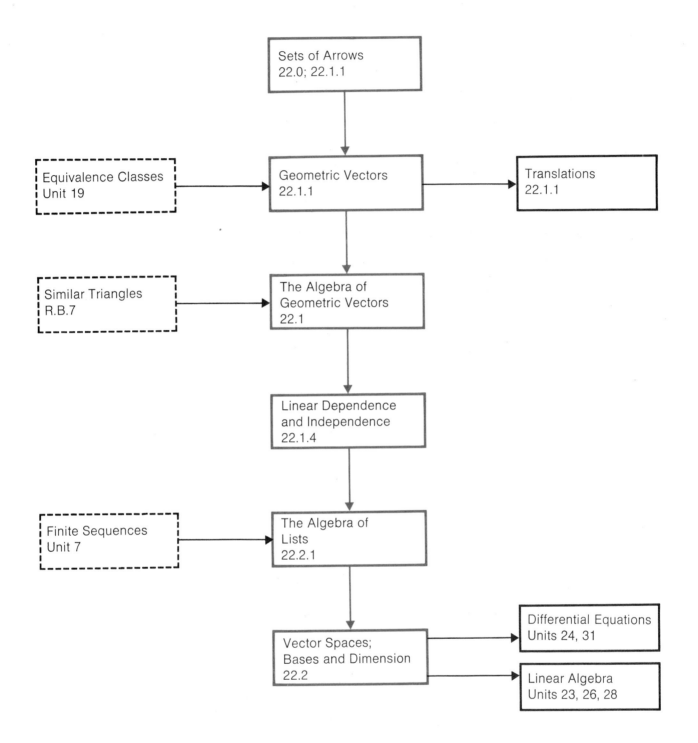

Glossary

Terms which are defined in this glossary are printed in CAPITALS.

ARROW	An ARROW is defined by a length, a direction and an end-point. It is associated with the particular point in space to which its tail end is attached.	3
BASE VECTORS	If $\{\underline{v}_1, \underline{v}_2, \ldots, \underline{v}_n\}$ is a LINEARLY INDEPENDENT SET of VECTORS of a VECTOR SPACE V, and this set also SPANS V, then it is called a set of BASE VECTORS of V.	23, 45
BASIS	A BASIS of a VECTOR SPACE V is a set of BASE VECTORS of V.	23, 45
DIMENSION	The DIMENSION of a VECTOR SPACE V is the (unique) number of elements in a BASIS of V. (If a basis does not consist of a *finite* number of elements, the vector space is said to be *infinite–dimensional*.)	45
DOT PRODUCT	See INNER PRODUCT.	
GEOMETRIC VECTOR	A GEOMETRIC VECTOR is the set of all ARROWS which have the same length and direction as some given arrow.	3
INNER PRODUCT	The INNER PRODUCT (SCALAR or DOT PRODUCT) of two GEOMETRIC VECTORS is the product of the lengths of the vectors and the cosine of the angle between them.	28
LINEAR COMBINATION	A LINEAR COMBINATION of the elements $\underline{v}_1, \underline{v}_2, \ldots, \underline{v}_k$ of a VECTOR SPACE V is an expression of the form $\alpha_1 \underline{v}_1 + \alpha_2 \underline{v}_2 + \cdots + \alpha_k \underline{v}_k$, where $\alpha_1, \alpha_2, \ldots, \alpha_k$ are SCALARS.	20, 43
LINEARLY DEPENDENT	The set of VECTORS $\{\underline{v}_1, \underline{v}_2, \ldots, \underline{v}_k\}$ of a VECTOR SPACE V is LINEARLY DEPENDENT if and only if there exist SCALARS $\alpha_1, \alpha_2, \ldots, \alpha_k$, which are not all zero, such that $\alpha_1 \underline{v}_1 + \alpha_2 \underline{v}_2 + \cdots + \alpha_k \underline{v}_k = \underline{0}$.	22, 43
LINEARLY INDEPENDENT	The set of VECTORS $\{\underline{v}_1, \underline{v}_2, \ldots, \underline{v}_k\}$ of a VECTOR SPACE V is LINEARLY INDEPENDENT if and only if $\alpha_1 \underline{v}_1 + \alpha_2 \underline{v}_2 + \cdots + \alpha_k \underline{v}_k = \underline{0}$ $\Rightarrow \alpha_1 = \alpha_2 = \cdots = \alpha_k = 0.$	22, 43
LIST	A LIST is a finite sequence of numbers, written either as $(a_1, a_2, \ldots, a_n)$ or as $\begin{pmatrix} a_1 \\ a_2 \\ \cdot \\ \cdot \\ \cdot \\ a_n \end{pmatrix}$	32
MODULUS OF A GEOMETRIC VECTOR	The MODULUS OF A GEOMETRIC VECTOR is the (positive) length of the vector.	15
SCALAR	In this text a SCALAR is a real number.	17
SCALAR PRODUCT	See INNER PRODUCT.	
SPAN	The set of VECTORS $\{\underline{v}_1, \underline{v}_2, \ldots, \underline{v}_n\}$ of a VECTOR SPACE V is said to SPAN V if each element of V can be expressed as a LINEAR COMBINATION of $\underline{v}_1, \underline{v}_2, \ldots, \underline{v}_n$.	20, 45

Notation

The symbols are presented in the order in which they appear in the text.

Bibliography

Banesh Hoffmann, *About Vectors*, (Prentice-Hall, 1966). This is an excellent book which discusses in detail the difficulties involved in defining *vectors* sensibly. It is very suitable for students who have met vectors before and are unhappy with our treatment of the subject. The notation and treatment used in the book are different from ours, but it will explain to some extent why we have chosen the particular approach adopted in the Foundation Course.

A. J. Pettofrezzo, *Vectors and Their Applications*, (Prentice-Hall, 1966). This is a straightforward treatment of vectors defined as "directed line segments"; it is suitable for students who find the equivalence class definition confusing.

Hans Liebeck, *Algebra for Scientists and Engineers*, (John Wiley, 1969). This book is an excellent introduction to Linear Algebra. The approach is similar to that of the Foundation Course.

Acknowledgement

Grateful acknowledgement is made to the following source for material used in this unit:

Prentice-Hall Inc. for Banesh Hoffmann, *About Vectors*.

"The Singular Incident of the Vectorial Tribe.

It is rumored that there was once a tribe of Indians who believed that arrows are vectors. To shoot a deer due northeast, they did not aim an arrow in the northeasterly direction; they sent two arrows simultaneously, one due north and the other due east, relying on the powerful resultant of the two arrows to kill the deer.

Skeptical scientists have doubted the truth of this rumor, pointing out that not the slightest trace of the tribe has ever been found. But the complete disappearance of the tribe through starvation is precisely what one would expect under the circumstances; and since the theory that the tribe existed confirms two such diverse things as the NONVECTORIAL BEHAVIOR OF ARROWS and the DARWINIAN PRINCIPLE OF NATURAL SELECTION, it is surely not a theory to be dismissed lightly."

BANESH HOFFMANN
About Vectors
(Prentice-Hall, 1966)

22.0 INTRODUCTION

At the beginning of our Foundation Course we introduced the basic concept of a set. When we introduce an operation which allows us to combine elements of a set (such as addition on the set of real numbers), then we have a mathematical structure and things become more interesting. With two operations (such as addition and multiplication) the structure becomes even more interesting, and more so when we introduce mappings from the original set to another set. At each stage we make the structure more intricate, and possibly more intriguing. At some stage it may be possible to prove results which are not obvious, and indeed, results which are completely unexpected. These results may also have useful applications.

Our intention in this unit is to build a particular mathematical structure. We begin with a set of elements which we call *arrows*; we then use arrows to define the set of *geometric vectors* and we discuss ways of combining geometric vectors. Even if this were simply an exercise in mathematical construction, it would be worth the time devoted to it. However, the system which we construct will lead us eventually to the subject of Linear Algebra, which is an important topic in mathematics and is the central theme in one of our second level courses.

The most important concept of this text is that of a *vector space*; there will be three more units, under the title *Linear Algebra*, which will demonstrate the importance of this concept. Once again we are beginning a new subject and again it is rather hard to appreciate the full significance of some of the material during the initial stages.

The first part of this text is devoted to geometric vectors for three reasons: you may have met this example of a vector space before; it is a geometric example of a vector space and many students find a pictorial approach to mathematics helpful; this geometric example is often an aid to intuition when dealing with vector spaces which are not geometric in character.

The fact that geometric vectors are extremely useful in applied mathematics is something of a red herring in this context. However, we give a short appendix which attempts to explain briefly how geometric vectors are used, and you may find it helpful to refer to it.

In section 22.2 we make what is almost a new start on the subject of vector spaces from a different point of view, and if you find the geometric vector approach hard to grasp, then you have in this section a fresh opportunity to understand the essential notion of a vector space.

If you have met vectors before, then we advise you to notice particularly our terminology. We call the vectors which commonly arise in applied mathematics (sometimes defined as "directed line segments") *geometric vectors*, and we use the word *vector* for an element of what we call a *vector space*.

22.1 GEOMETRIC VECTORS

22.1.0 Introduction

We all know that the speed of an aeroplane relative to the ground is affected by the speed of the wind. With the wind behind it, it flies faster; in a cross-wind the pilot must aim the aircraft slightly into the wind in order to reach his destination. The applied mathematician often has to deal with a situation like this in which he needs to make a mathematical model of the physical situation.

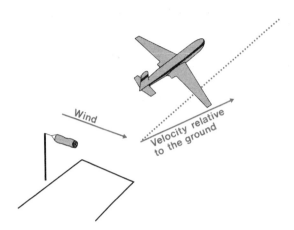

He assumes that the aircraft is being moved bodily along with the wind, just as a puff of smoke might be, and that he can represent this motion by an arrow, the length representing the wind speed and the direction representing the wind direction. For example, there may be a 30 mile/h easterly wind blowing over the country.

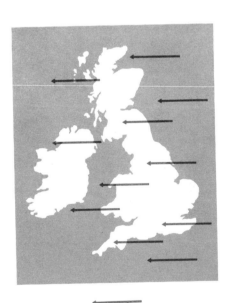

(Incidentally he also assumes that the aircraft is compressed to a point, since its shape and size is not important for the problem under investigation.) The speed of the aircraft in the moving air stream (the air speed) can be represented by another arrow. We shall see that the way we can combine these arrows mathematically, by considering them as representatives of geometric vectors, enables the applied mathematician to determine the direction in which the aircraft should point.

22.1.1   Sets of Arrows and Geometric Vectors

In this section we introduce *geometric vectors* using the concept of an *arrow*.

Arrows

As far as we are concerned, an arrow has length, direction, and position. The position can be specified by one end-point, the blunt end, say.

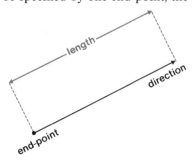

Sometimes we restrict all the arrows to lie in a plane (in other words, they can all be drawn on a flat piece of paper). On other occasions they can be in three-dimensional space but, unless it is clear from the context, we shall have to declare which case we mean at the outset of a discussion. For the moment everything we say applies to both cases.

Our first requirement is a notation, and we shall use a small arrow over a letter, for example $\vec{a}$, to indicate that the element denotes an arrow. Sometimes an arrow is defined in terms of its end-points, say A and B, in which case we can write $\overrightarrow{AB}$ for the arrow with its blunt end at A and sharp end at B.

We define two arrows to be equal if they have the same length, direction and position. It is important to notice that the two arrows $\overrightarrow{AB}$ and $\overrightarrow{PQ}$ shown below are distinct, even though they have the same length and direction. (They may, for example, represent the velocities associated with distinct particles at the points A and P.)

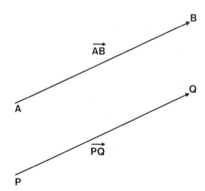

Geometric Vectors

The set of *all* arrows with the same length and direction as some given arrow, but with different positions, is of interest to us because sometimes we wish to impart information in bulk, as it were, rather than for each individual arrow. (For example, we may wish to represent "a force 6 easterly wind".) We call such a set a geometric vector. On page 4 we have illustrated some of the arrows belonging to the same geometric vector. Remember that a geometric vector is the set of *all* arrows with the same length and direction as some given arrow, and although each of the arrows in the diagram is distinct, any one of them will determine the same geometric vector.

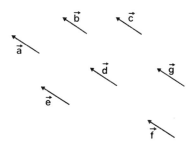

We need a notation for geometric vectors. Since any one arrow, say $\overrightarrow{AB}$, is sufficient to specify the geometric vector to which it belongs, we choose to denote that geometric vector by $\underrightarrow{AB}$. Since geometric vectors are sets of arrows, the equality of two geometric vectors is defined: every arrow belonging to one set must also belong to the other. The two arrows $\overrightarrow{AB}$ and $\overrightarrow{PQ}$ belong to the same geometric vector, which we can denote by $\underrightarrow{AB}$ or $\underrightarrow{PQ}$; these geometric vectors are equal, so we write $\underrightarrow{AB} = \underrightarrow{PQ}$.

Notation 2
* * *

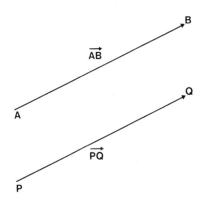

We use a similar notation, $\underrightarrow{a}$, for the geometric vector of which $\vec{a}$ is a member; that is, $\vec{a} \in \underrightarrow{a}$.

It is impossible to draw a picture which includes *every* arrow belonging to a geometric vector, so when we need a pictorial representation of a geometric vector we draw just one of its arrows, with the understanding that it is a representative from the set.

The set of all geometric vectors is the starting point for the construction of our algebra of vectors. Before we can usefully combine geometric vectors, we need to ensure that the particular arrows chosen as representatives do not affect the result. Let us look at some aspects of geometric vectors which relate the concept to some other ideas which we met in *Unit 19*.

Equivalence Classes

The relation:
        has the same length and direction as
is an equivalence relation on the set of all arrows. (For the definition of *equivalence relation* see *Unit 19, Relations*.)

If we abbreviate

        $\vec{a}$ has the same length and direction as $\vec{b}$

to

        $\vec{a} \, \rho \, \vec{b}$

then we have:

(i) $\vec{a} \, \rho \, \vec{a}$ for all $\vec{a}$;
(ii) $\vec{a} \, \rho \, \vec{b} \Rightarrow \vec{b} \, \rho \, \vec{a}$;
(iii) $\vec{a} \, \rho \, \vec{b}$ and $\vec{b} \, \rho \, \vec{c} \Rightarrow \vec{a} \, \rho \, \vec{c}$;

that is, the three requirements of an equivalence relation are satisfied. This means that $\underline{a}$ is the equivalence class which contains the element $\vec{a}$.

An equivalence relation partitions the original set (here, the set of all arrows) into non-overlapping subsets, called equivalence classes (here, the geometric vectors): it is this fact which makes equivalence classes important. In other words, any two geometric vectors $\underline{a}$ and $\underline{b}$ are either identical or they do not have any element in common.

Translations

In a plane, we can interpret the geometric vector $\underline{a}$, comprising arrows lying in the plane, as a command to each point of the plane to move to a new position. For example, a particular point P goes to the point Q if $\overrightarrow{PQ} \in \underline{a}$.

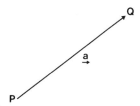

This is very reminiscent of the idea of a mapping in which Q is the image of P, and in fact this is sometimes a helpful way of looking at geometric vectors. We can use the geometric vector $\underline{a}$ to define a one-one function f, with domain and codomain the set of points in the plane, such that

$$f : P \longmapsto Q$$

where $PQ = \underline{a}$. Such a function is called a translation. To each geometric vector there corresponds a unique translation.

Definition 4

In a similar way we could define a translation of three-dimensional space.

Exercise 1

Exercise 1
(2 minutes)

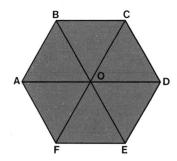

The above figure is a regular hexagon. We have omitted the arrow heads as these are implied in the following statements. In each case indicate if the statement is true or false:

(i) $\overrightarrow{AB} = \overrightarrow{ED}$ TRUE/FALSE

(ii) $\overrightarrow{AB} = \underrightarrow{AB}$ TRUE/FALSE

(iii) $\underrightarrow{FO} = \underrightarrow{ED}$ TRUE/FALSE

(iv) $\underrightarrow{AO} = \underrightarrow{EF}$ TRUE/FALSE

(v) $\underrightarrow{BC} = \underrightarrow{AD}$ TRUE/FALSE

■

Solution 1

   (i) FALSE. Equal arrows must have the same length, direction and *position.*

   (ii) FALSE. $\overrightarrow{AB}$ is an arrow, whereas $\underset{\longrightarrow}{AB}$ is a set of arrows.

   (iii) TRUE. Both the arrows $\overrightarrow{FO}$ and $\overrightarrow{ED}$ belong to the same geometric vector.

   (iv) FALSE. The arrows $\overrightarrow{AO}$ and $\overrightarrow{EF}$ (representatives of $\underset{\longrightarrow}{AO}$ and $\underset{\longrightarrow}{EF}$ respectively) are in opposite directions.

   (v) FALSE. The arrows $\overrightarrow{BC}$ and $\overrightarrow{AD}$ have different lengths. ■

22.1.2   Addition on the Set of Geometric Vectors

We have our set of geometric vectors, and the problem is now to combine them in a mathematically interesting fashion. We might ask what sort of combinations will be useful, but since for the moment we are regarding this as a mathematical exercise rather than as an attack on an external problem, that question is not really relevant. If we find (as, of course, we do) that our mathematical system has some useful applications, so much the better, but that is not our prime concern at present.

Discussion
*

Addition of Geometric Vectors

First we shall define an operation + which we shall call *addition* of geometric vectors; *addition* is a good word to use because the operation has very similar properties to the operation of addition on *R*.

Main Text
* * *

For any two geometric vectors $\underset{\sim}{a}$ and $\underset{\sim}{b}$ we define $\underset{\sim}{a} + \underset{\sim}{b}$ as follows. Take any arrow $\vec{a} \in \underset{\sim}{a}$, then choose the arrow $\vec{b} \in \underset{\sim}{b}$ which has its blunt end at the sharp end of $\vec{a}$.

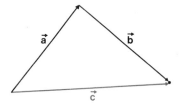

The arrow $\vec{c}$, with its blunt end at the blunt end of $\vec{a}$ and sharp end at the sharp end of $\vec{b}$, belongs to some geometric vector, $\underset{\sim}{c}$, and we define this geometric vector to be $\underset{\sim}{a} + \underset{\sim}{b}$.

Definition 1
* * *

Have we really defined a binary operation on the set of geometric vectors? One of the requirements of a binary operation is that it should give a *unique* answer. We formed $\underset{\sim}{a} + \underset{\sim}{b} = \underset{\sim}{c}$ by taking *any* $\vec{a} \in \underset{\sim}{a}$, which determined $\vec{b} \in \underset{\sim}{b}$, and then combining $\vec{a}$ and $\vec{b}$ to give $\underset{\sim}{c}$. Suppose we take a different representative $\vec{a}_1 \in \underset{\sim}{a}$; will we still get the same $\underset{\sim}{c}$?

Discussion
* *

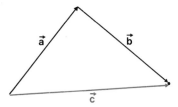 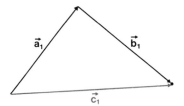

The two triangles shown are congruent, and it follows that $\vec{c}$ and $\vec{c}_1$ have the same length: the triangles are also similar, and therefore $\vec{c}$ is parallel to $\vec{c}_1$, so $\vec{c}$ and $\vec{c}_1$ have the same direction. So both $\vec{c}$ and $\vec{c}_1$ belong to $\underset{\sim}{c}$. The geometric vector $\underset{\sim}{c}$ is therefore uniquely defined.

(See RB7)

We can represent $\underset{\sim}{a} + \underset{\sim}{b}$ by the following diagram. (The arrows in this diagram are merely representatives from the corresponding geometric vectors; we have labelled them as geometric vectors.)

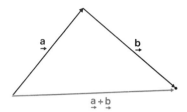

(We can also look at this using the ideas of *Unit 19, Relations*. We have already seen that a geometric vector is an equivalence class of arrows. We can define a pseudo-addition on the set of arrows for which we can add $\vec{a}$ to $\vec{b}$ to give $\vec{c}$ as defined above. (It is not a proper addition, since we cannot "add" *every* pair of arrows unless we modify our definition of addition.) We also have the natural mapping of the set of arrows to the set of geometric vectors defined by

$$\vec{a} \longmapsto \underset{\sim}{a}$$

and we can ask whether this is compatible with the pseudo-addition. The argument is then similar to the above, and we find that we can define the induced binary operation of addition on the set of geometric vectors.)

Discussion
*

Exercise 1

Draw a diagram to illustrate the geometric vectors $\underset{\sim}{a} + \underset{\sim}{b}$ and $\underset{\sim}{b} + \underset{\sim}{a}$. Are the two geometric vectors equal? Is addition of geometric vectors commutative? ■

Exercise 1
(2 minutes)

Exercise 2

Use the following diagram to illustrate the associative property of addition of geometric vectors:

$$(\underset{\sim}{a} + \underset{\sim}{b}) + \underset{\sim}{c} = \underset{\sim}{a} + (\underset{\sim}{b} + \underset{\sim}{c})$$

Exercise 2
(2 minutes)

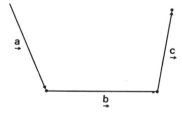

Exercise 3

We saw on page 5 that a geometric vector determines a unique translation which maps the set of points in the plane (or three-dimensional space) to itself. So the set of geometric vectors with addition will determine the set of translations with a binary operation. What is the binary operation on translations corresponding to addition on geometric vectors? ■

Exercise 3
(3 minutes)

Solution 1   **Solution 1**

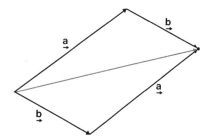

It is true that $\underset{\sim}{a} + \underset{\sim}{b} = \underset{\sim}{b} + \underset{\sim}{a}$, and therefore addition is commutative.   ■

Solution 2   **Solution 2**

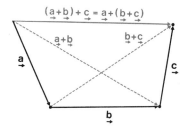

■

Solution 3   **Solution 3**

The required binary operation is composition on the set of translations (i.e. perform one translation, and then the other). We can think of the set of geometric vectors being mapped on to the set of translations by a one-one function m:

$$m : \underset{\sim}{a} \longmapsto f$$

This function m is then an *isomorphism* of the set of geometric vectors with addition to the set of translations with composition.

$$
\begin{array}{ccc}
(\underset{\sim}{a}, \underset{\sim}{b}) & \overset{+}{\longrightarrow} & \underset{\sim}{a} + \underset{\sim}{b} \\
m \downarrow & & \downarrow m \\
(f, g) & \overset{\circ}{\longrightarrow} & f \circ g = g \circ f
\end{array}
$$

Notice that, although composition of functions is not normally commutative, we know that it is commutative for the set of translations, because it corresponds to the addition of geometric vectors, which is commutative.
■

The Zero Geometric Vector

Is the binary operation of addition on the set of geometric vectors closed?   Main Text
* *
Before answering this question, we give a further definition.

Given a geometric vector $\underset{\sim}{a}$, we define $-\underset{\sim}{a}$ to be the geometric vector   Definition 2
* * *
determined by the arrow with the same length but the opposite direction
to $\vec{a}$, where $\vec{a} \in \underset{\sim}{a}$.

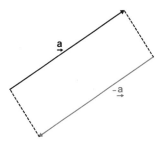

8

What happens if we add $\underset{\rightarrow}{a}$ to $-\underset{\rightarrow}{a}$?

We have defined geometric vectors in terms of arrows, which have length and direction; this leads to a slight problem with the "zero element". A zero length is all right, but what is its direction? Because this question has no satisfactory answer, we begin our process of abstraction and define the zero geometric vector as the result of adding *any* two geometric vectors of the form $\underset{\rightarrow}{a}$ and $-\underset{\rightarrow}{a}$. Notice that $\underset{\rightarrow}{a} + (-\underset{\rightarrow}{a})$ and $\underset{\rightarrow}{b} + (-\underset{\rightarrow}{b})$ define the same zero geometric vector, which we denote by $\underset{\rightarrow}{0}$. (We are saying that, strictly speaking, $\underset{\rightarrow}{0}$ is not a geometric vector, as it has no direction, but it is convenient to *call* it a geometric vector. In much the same way, the number zero in R is not really on a par with the other real numbers.) By analogy with zero in R, we define $-\underset{\rightarrow}{0} = \underset{\rightarrow}{0}$ (since $\underset{\rightarrow}{0}$ has zero length, we cannot "turn it round").

Definition 3
* * *

Notation 1
* * *

Definition 4
* * *

It follows from Definition 3 and the associative property of $+$ that

$$\underset{\rightarrow}{a} + \underset{\rightarrow}{0} = \underset{\rightarrow}{0} + \underset{\rightarrow}{a} = \underset{\rightarrow}{a}$$

(cf. $a + 0 = 0 + a = a$ where $a \in R$). The element $\underset{\rightarrow}{0}$ corresponds to the translation which maps each point to itself, i.e. for which nothing moves.

With the inclusion of the zero geometric vector, our set of geometric vectors with addition is now closed.

The Operation of Subtraction

Let us summarize our present position. We have the set of geometric vectors, with the closed binary operation of addition defined on it. We also have a zero element $\underset{\rightarrow}{0}$ in the set. What about subtraction?

Discussion

There is a difference in the real number system between

the element which when added to $+2$ gives zero,

that is,

the signed integer -2

and

the instruction: "subtract the number $+2$",

which we may write as

$-(+2)$.

In the first case the $-$ is a label attached to the 2 to show that the number is negative: in the second case the $-$ denotes the binary operation of subtraction. For geometric vectors, we have the equivalent of the first case, but we do not yet have the equivalent of the second case as we have not defined *subtraction* of geometric vectors.

In the real number system, there is a connection between the signed number -2 and "subtract $+2$": "subtract $+2$" is equivalent to "add -2".

By analogy, we now define

$\underset{\rightarrow}{a} - \underset{\rightarrow}{b}$ to be $\underset{\rightarrow}{a} + (-\underset{\rightarrow}{b})$.

Definition 5
* * *

This may seem obvious, but it is an essential step. The result of subtracting $\underset{\sim}{b}$ from $\underset{\sim}{a}$ is shown below.

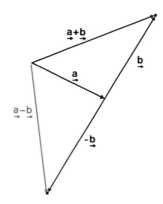

Notice that the operation $-$ is a binary operation. From Definition 2 it follows that

$$-(-\underset{\sim}{b}) = \underset{\sim}{b}$$

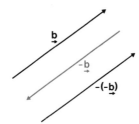

So

$$\underset{\sim}{a} - (-\underset{\sim}{b}) = \underset{\sim}{a} + (-(-\underset{\sim}{b}))$$
$$= \underset{\sim}{a} + \underset{\sim}{b}.$$

Properties of Geometric Vectors

Summary
* * *

(i) $\underset{\sim}{a} + \underset{\sim}{b}$ is a geometric vector ($+$ is closed).
(ii) $\underset{\sim}{a} + (\underset{\sim}{b} + \underset{\sim}{c}) = (\underset{\sim}{a} + \underset{\sim}{b}) + \underset{\sim}{c}$ ($+$ is associative).
(iii) $\underset{\sim}{a} + \underset{\sim}{b} = \underset{\sim}{b} + \underset{\sim}{a}$ ($+$ is commutative).
(iv) To each geometric vector $\underset{\sim}{a}$ there corresponds a unique geometric vector $-\underset{\sim}{a}$.
(v) There is a geometric vector $\underset{\sim}{0}$ such that for all $\underset{\sim}{a}$

$$\underset{\sim}{a} + (-\underset{\sim}{a}) = (-\underset{\sim}{a}) + \underset{\sim}{a} = \underset{\sim}{0}$$

and

$$\underset{\sim}{a} + \underset{\sim}{0} = \underset{\sim}{0} + \underset{\sim}{a} = \underset{\sim}{a}$$

(vi) $-\underset{\sim}{0} = \underset{\sim}{0}$
(vii) Subtraction of geometric vectors is defined by

$$\underset{\sim}{a} - \underset{\sim}{b} = \underset{\sim}{a} + (-\underset{\sim}{b})$$

Exercise 4

Exercise 4
(3 minutes)

From the definitions, prove that

$$(\underset{\sim}{a} - \underset{\sim}{b} = \underset{\sim}{c}) \Rightarrow (\underset{\sim}{a} = \underset{\sim}{c} + \underset{\sim}{b})$$

■

Exercise 5

Let the points O, A, B, C and D in the xy-plane have co-ordinates $(0, 0)$, $(-2, 1)$, $(3, 2)$, $(2, 4)$ and $(1, 5)$ respectively. Construct representatives of the following geometric vectors graphically.

(i) $\overrightarrow{OA} + \overrightarrow{OB}$

(ii) $\overrightarrow{OB} - \overrightarrow{BC}$

(iii) $\overrightarrow{OB} + \overrightarrow{BC} + \overrightarrow{CD} + \overrightarrow{DA} + \overrightarrow{AO}$

(Remember that each of the arrows which you draw represents a geometric vector and can therefore be moved to any position on the paper provided that the length and direction are kept the same.) ■

Solution 4

$$(\underline{a} - \underline{b} = \underline{c}) \Rightarrow \underline{a} + (-\underline{b}) = \underline{c} \qquad \text{(definition of subtraction)}$$

$$\Rightarrow (\underline{a} + (-\underline{b})) + \underline{b} = \underline{c} + \underline{b}$$

$$\Rightarrow \underline{a} + ((-\underline{b}) + \underline{b}) = \underline{c} + \underline{b} \qquad \text{(associativity of } +)$$

$$\Rightarrow \underline{a} + \underline{0} = \underline{c} + \underline{b} \qquad \text{(definition of } \underline{0})$$

$$\Rightarrow \underline{a} = \underline{c} + \underline{b} \qquad \text{(property of } \underline{0}) \qquad \blacksquare$$

Solution 5

(i) $\overrightarrow{OA}$ and $\overrightarrow{OB}$ are the two arrows shown below in red.

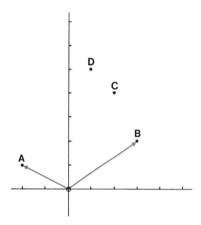

Our problem is to find $\underline{OA} + \underline{OB}$.

We can choose $\overrightarrow{OA}$ as the representative of the geometric vector $\underline{OA}$; then we need the arrow $\overrightarrow{AP}$ as the representative of $\underline{OB}$.

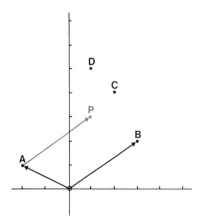

We can now see that

$$\underline{OA} + \underline{OB} = \underline{OA} + \underline{AP} = \underline{OP}$$

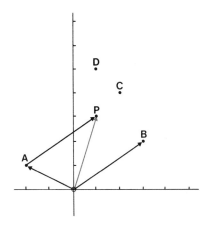

(ii) In the following diagram, $-\underrightarrow{BC} = \underrightarrow{BQ}$ and therefore

$$\underrightarrow{OB} - \underrightarrow{BC} = \underrightarrow{OB} + \underrightarrow{BQ} = \underrightarrow{OQ}$$

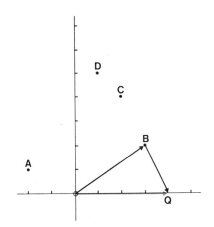

(iii) From the diagram below we can see that

$$\underrightarrow{OB} + \underrightarrow{BC} = \underrightarrow{OC}$$

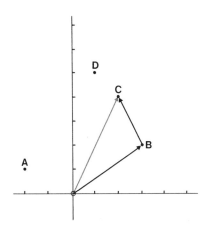

Then $\underrightarrow{OC} + \underrightarrow{CD} = \underrightarrow{OD}$,

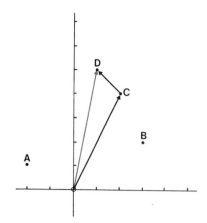

also $\underrightarrow{OD} + \underrightarrow{DA} = \underrightarrow{OA}$,

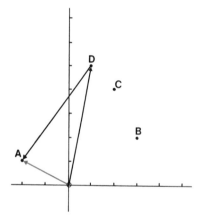

but $\underrightarrow{OA} + \underrightarrow{AO} = \underrightarrow{0}$.

It follows that

$$\underrightarrow{OB} + \underrightarrow{BC} + \underrightarrow{CD} + \underrightarrow{DA} + \underrightarrow{AO} = \underrightarrow{0}$$

22.1.3 Scalar Multiples of Geometric Vectors

In this section we consider lengths and scalar multiples of geometric vectors.

Length of a Geometric Vector

It is convenient to have a notation for the length of a geometric vector, and we denote the length of a by $|a|$, which we read as "the modulus of a". We have used the word *modulus* previously for real numbers (e.g. $|-2| = 2$), and we shall use it again for complex numbers. In each case where we use *modulus*, we are considering the magnitude of the quantity only, ignoring "direction". (For example, on the real number line, -2 and 2 are the same distance from 0, but in opposite directions.) We write the length of $\overrightarrow{AB}$ as either $|\overrightarrow{AB}|$ or AB.

Suppose that

$$c = a + b$$

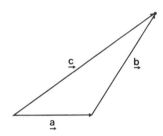

The length of any side of a triangle is less than (or equal to) the sum of the other two.* In the modulus notation this statement is

$$|c| \leqslant |a| + |b|$$

This inequality is called the triangle inequality.

It is only possible to have equality when the vertices of the triangle lie in a straight line, and then the interior of the triangle disappears completely.

Example 1

Example 1

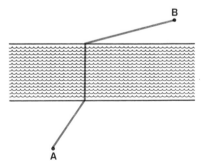

A (mythical) man wishes to travel regularly from A to B, and he wants to build a bridge across the river at right-angles to the bank (to keep the bridge as short as possible) in the position which will minimize his travelling distance. Where should he put the bridge? ■

---

* This is a theorem in Euclidean geometry.

Solution of Example 1

It is a waste of time using calculus on this problem: there are easier ways of doing it. Consider the following diagram.

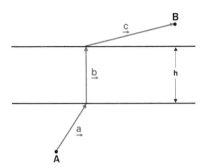

The geometric vectors represent the three distinct parts of the man's journey, and clearly he needs to walk in a straight line on each bank. Our problem is to choose the position of the bridge which will minimize

$$|\underrightarrow{a}| + |\underrightarrow{b}| + |\underrightarrow{c}|.$$

The value of $|\underrightarrow{b}|$ is h and is not affected by the position of the bridge, so our problem is to minimize $|\underrightarrow{a}| + |\underrightarrow{c}|$. We know from the triangle inequality that

$$|\underrightarrow{a} + \underrightarrow{c}| \leqslant |\underrightarrow{a}| + |\underrightarrow{c}|,$$

and we only get equality when $\underrightarrow{a}$ and $\underrightarrow{c}$ are parallel.* Hence, the minimum value occurs when $\underrightarrow{a}$ and $\underrightarrow{c}$ are parallel. It only remains to use this information to find the position of the bridge. ■

Exercise 1

How does the man determine the position of the bridge to achieve the minimum distance? ■

Exercise 1
(3 minutes)

Multiplication of a Geometric Vector by a Scalar

We know what we mean by $\underrightarrow{a} + \underrightarrow{a}$, but if we were to abbreviate this to $2\underrightarrow{a}$ without further ado, we would be missing a point.

Main Text
* * *

It seems natural to write

$$\underrightarrow{a} + \underrightarrow{a} = 2\underrightarrow{a},$$

but we should not do so without explaining what we mean by $2\underrightarrow{a}$. The definition is thrust upon us, because $\underrightarrow{a} + \underrightarrow{a}$ is in the same direction as $\underrightarrow{a}$ but has twice its length.

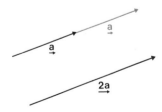

* The geometric vectors $\underrightarrow{a}$ and $\underrightarrow{c}$ are defined to be parallel if the arrows $\vec{a}$ and $\vec{c}$ are parallel, where $\vec{a} \in \underrightarrow{a}$ and $\vec{c} \in \underrightarrow{c}$.

For $\lambda \in R$ we define $\lambda\underset{\sim}{a}$ ($\lambda > 0$) to be a geometric vector in the same direction as $\underset{\sim}{a}$ but with length $\lambda|\underset{\sim}{a}|$. This implies that $|\lambda\underset{\sim}{a}| = \lambda|\underset{\sim}{a}|$. If $\lambda < 0$, then we define $\lambda\underset{\sim}{a}$ to be in the opposite direction to $\underset{\sim}{a}$, with length $-\lambda|\underset{\sim}{a}|$. We define $0\underset{\sim}{a} = \underset{\sim}{0}$. So

$$|\lambda\underset{\sim}{a}| = |\lambda| \times |\underset{\sim}{a}|.$$

Definition 2
* * *

When we write $\lambda\underset{\sim}{a}$, we say that the geometric vector $\underset{\sim}{a}$ is *multiplied* by the *scalar λ*. Although λ is just a real number in the present context, we call it a scalar to distinguish it from a geometric vector, because there exist more general situations in which λ is a scalar but not a real number.

Definition 3
* * *

Further Properties of Geometric Vectors

Using our definition of $\lambda\underset{\sim}{a}$, you should be able to prove the following results, which are in accord with our intuitive expectations:

Summary
* * *

(viii) when $\lambda = 0$, we have $0\underset{\sim}{a} = \underset{\sim}{0}$;

 (ix) when $\lambda = 1$, we have $1\underset{\sim}{a} = \underset{\sim}{a}$;

  (x) when $\lambda = -1$, we have $-1\underset{\sim}{a} = -\underset{\sim}{a}$;

 (xi) for any geometric vector $\underset{\sim}{a}$ and any real number λ, $\lambda\underset{\sim}{a}$ is a geometric vector;

(xii) $\lambda(\underset{\sim}{a} + \underset{\sim}{b}) = \lambda\underset{\sim}{a} + \lambda\underset{\sim}{b}$ for any geometric vectors $\underset{\sim}{a}$ and $\underset{\sim}{b}$ and any real number λ (see the following exercise);

(xiii) for any geometric vector $\underset{\sim}{a}$ and any real numbers λ and μ,

$$(\lambda + \mu)\underset{\sim}{a} = \lambda\underset{\sim}{a} + \mu\underset{\sim}{a};$$

(xiv) for any geometric vector $\underset{\sim}{a}$ and any real numbers λ and μ,

$$(\lambda\mu)\underset{\sim}{a} = \lambda(\mu\underset{\sim}{a}).$$

Exercise 2

Exercise 2
(3 minutes)

Demonstrate property (xii), i.e. that the function $\underset{\sim}{a} \longmapsto \lambda\underset{\sim}{a}$ is a morphism of the set of geometric vectors under addition to itself:

$$\lambda(\underset{\sim}{a} + \underset{\sim}{b}) = \lambda\underset{\sim}{a} + \lambda\underset{\sim}{b} \qquad\qquad \blacksquare$$

In the next section we shall use our simple algebraic system to develop the powerful concept of *linear dependence*.

Discussion

The next exercise serves as an introduction to the following section, in which we shall be interested in the following questions:

 (i) If we are given two geometric vectors $\underset{\sim}{a}$ and $\underset{\sim}{b}$, can we add suitably chosen scalar multiples of them to produce a third given geometric vector $\underset{\sim}{c}$?

(ii) If we produce $\underset{\sim}{c}$ as a sum of scalar multiples of $\underset{\sim}{a}$ and $\underset{\sim}{b}$, is it possible to do it in several distinct ways?

Solution 1   **Solution 1**

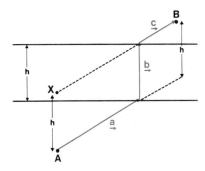

When drawing his plan, he begins at A by drawing in the arrow $\overrightarrow{AX}$, which has the same length and direction as the bridge, and then joins X to B. Where XB cuts the far bank is the required position for the other end of the bridge.   ■

Solution 2   **Solution 2**

If $\lambda = 0$, there is not much to demonstrate. Suppose $\lambda > 0$; then we have the following diagram, where $\underline{c} = \lambda\underline{a} + \lambda\underline{b}$. We want to show that $\underline{c} = \lambda(\underline{a} + \underline{b})$.

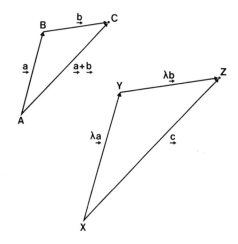

Since $\dfrac{XY}{AB} = \dfrac{YZ}{BC} = \lambda$, and angle $A\widehat{B}C = $ angle $X\widehat{Y}Z$, the two triangles are similar. Therefore   (See RB7)

$$\frac{XZ}{AC} = \lambda$$

Further, XZ is parallel to AC, and therefore $\underline{XZ} = \lambda\underline{AC}$, i.e.

$$\underline{c} = \lambda(\underline{a} + \underline{b})$$

If $\lambda < 0$, we have a similar argument.   ■

Exercise 3

If a and b are the two geometric vectors represented in the following diagram, mark the point determined as the image of the origin O under the translation corresponding to

$$\lambda a + \mu b,$$

where $\lambda = 2$ and $\mu = 3$. Label this point $(2, 3)$.

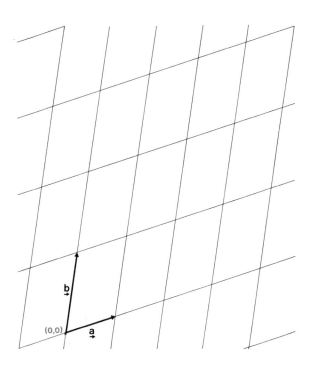

In general, if f is the translation corresponding to $\lambda a + \mu b$, and $f(0) = P$, then we label $P(\lambda, \mu)$.

Mark the points on the above diagram which have labels $(1, 1), (3, 2), (0, 1),$ $(0, 0), (-1, 1)$.

Now try the problem in reverse. Mark any point you like on the diagram, then find (approximate) values of λ and μ such that (λ, μ) is the label for your chosen point. ∎

19

Solution 3

Solution 3

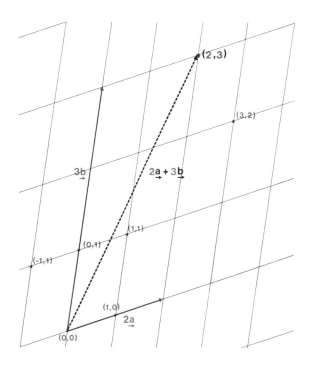

22.1.4   Linear Dependence and Independence

22.1.4

The previous exercise is reminiscent of the familiar rectangular Cartesian co-ordinate system. In fact, had we chosen a and b at right angles and of unit length, then the situation would be identical, and the pair (λ, μ) would be the co-ordinates of the point which is the image of the origin under the translation corresponding to $\lambda a + \mu b$. We call an expression such as $\lambda a + \mu b$ a linear combination of a and b.

Main Text
* * *

Definition 1
* * *

(To save ourselves having to repeat the phrase "the image of the origin under the translation corresponding to a" we shall write $a(0)$. This is an abuse of notation (a standard practice in mathematics) because a is not strictly speaking a function, although it can be used to define one, as we have seen.)

In our present context it does not seem to matter if a and b are at right angles or not. Given the origin, they can be used to determine any point we wish (in the plane of the paper) by making a suitable choice of λ and μ. We call the ordered pair of numbers (λ, μ) the co-ordinates of the point with respect to a, b and the point which has been chosen as the origin.

Definition 2
* * *

Suppose that we are given two geometric vectors a and b, and that it is possible to express *every* geometric vector in the plane of a and b as a linear combination of a and b. Then we say that $\{a, b\}$ spans the set of (plane) geometric vectors. This is equivalent to saying that, given any point of the plane, we can find two scalars (real numbers) λ and μ such that the expression

Definition 3
* * *

$$(\lambda a + \mu b)(0)$$

specifies that point.

(The values of λ and μ need not necessarily be integers or positive; they may be any real numbers. In other words, we are considering every point of the plane and not just those on the vertices of the grid as in Exercise 22.1.3.3.)

Will a set of any pair of geometric vectors span the set of (plane) geometric vectors? The answer is "No". To see this, suppose that we are given the pair of geometric vectors shown in the following diagram.

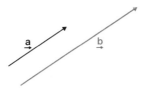

Whatever values of λ and μ we choose, the points $(\lambda \underline{a} + \mu \underline{b})(0)$ always lie in a straight line : we cannot get out of that line with this choice of geometric vectors. Essentially this is because $\underline{b}$ is a multiple of $\underline{a}$; in fact, $\underline{b} = 2\underline{a}$, so that

$$\lambda \underline{a} + \mu \underline{b} = \lambda \underline{a} + \mu(2\underline{a})$$
$$= (\lambda + 2\mu)\underline{a},$$

and every linear combination of $\underline{a}$ and $\underline{b}$ is simply a multiple of $\underline{a}$.

In this case $\{\underline{a}, \underline{b}\}$ only spans the set of geometric vectors parallel to $\underline{a}$, and this is because $\underline{b}$ contributes nothing new.

Things will go wrong (i.e. $\{\underline{a}, \underline{b}\}$ will not span the set of plane geometric vectors) if $\underline{b}$ is a multiple of $\underline{a}$, that is to say, if $\underline{b} = \beta \underline{a}$ for some real number β (including $\beta = 0$). Equally well, we are in trouble if $\underline{a}$ is a multiple of $\underline{b}$, so that $\underline{a} = \alpha \underline{b}$ for some real number α. This can be expressed in a symmetric form : things will go wrong if we can find two real numbers α_1 and α_2, which are *not both zero*, such that

$$\alpha_1 \underline{a} + \alpha_2 \underline{b} = \underline{0}$$

We say that $\{\underline{a}, \underline{b}\}$ is *linearly dependent* if and only if there are scalars α_1 and α_2, not both zero, which satisfy the above equation. We say that $\{\underline{a}, \underline{b}\}$ is *linearly independent* if it is not linearly dependent.

There is a simple, but important, consequence of linear independence. If we are told that $\{\underline{a}, \underline{b}\}$ is linearly independent, and yet there *are* scalars α_1 and α_2 such that

$$\alpha_1 \underline{a} + \alpha_2 \underline{b} = \underline{0}$$

then we can immediately deduce that $\alpha_1 = \alpha_2 = 0$.

Example 1

Example 1

We can show that $\{\underline{a}, 2\underline{a}\}$ is linearly dependent for any geometric vector $\underline{a}$ as follows.

We need to find two scalars α_1 and α_2, not both zero, such that

$$\alpha_1 \underline{a} + \alpha_2(2\underline{a}) = \underline{0}$$

Clearly $\alpha_1 = 2$ and $\alpha_2 = -1$ will do. (There are many other choices; for example, $\alpha_1 = 4$ and $\alpha_2 = -2$.) ■

We shall see later that the linear independence of $\{\underline{a}, \underline{b}\}$ is not only *necessary** but also *sufficient** to guarantee that it spans the set of (plane) geometric vectors.

Very similar arguments apply in three-dimensional space. We can extend the same idea to any number of geometric vectors to give the following definitions.

---

* The terms *necessary* and *sufficient* were defined in *Unit 17*.

The set of geometric vectors $\{\underline{a}^{(1)}, \underline{a}^{(2)}, \underline{a}^{(3)}, \dots, \underline{a}^{(n)}\}$ is said to be linearly dependent if and only if there are scalars $\alpha_1, \alpha_2, \dots, \alpha_n$, not all zero, such that

$$\alpha_1\underline{a}^{(1)} + \alpha_2\underline{a}^{(2)} + \alpha_3\underline{a}^{(3)} + \cdots + \alpha_n\underline{a}^{(n)} = \underline{0}.$$

Definition 4
* * *

(We shall also say that the geometric vectors themselves are linearly dependent.)

The set of geometric vectors $\{\underline{a}^{(1)}, \underline{a}^{(2)}, \dots, \underline{a}^{(n)}\}$ is said to be linearly independent if it is not linearly dependent. (We shall also say that the geometric vectors themselves are linearly independent.)

Definition 5
* * *

The essential feature about a linearly dependent set of geometric vectors is that it implies a certain amount of redundance in the set under consideration. If the set is linearly independent, there is no such redundance although it may still be "incomplete"; for example, a linearly independent set $\{\underline{a}, \underline{b}\}$ cannot span the whole set of geometric vectors in three dimensions. To make this intuitive idea clearer, let us look again at the two-dimensional problem, not this time in terms of two geometric vectors $\underline{a}$ and $\underline{b}$ lying in the plane, but in terms of three geometric vectors $\underline{a}, \underline{b}$ and $\underline{c}$ lying in the plane. There are two possible occurrences. Two or more of the geometric vectors are parallel, or they all have different directions (assuming that none of them is $\underline{0}$). Let us consider the two cases.

Discussion

(i) Suppose first that $\underline{a}$ and $\underline{b}$ are parallel, then $\{\underline{a}, \underline{b}\}$ is linearly dependent, because we can express one of the geometric vectors as a scalar multiple of the other: $\underline{a} = \lambda\underline{b}$ say.

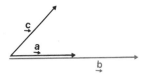

This means that $\underline{a}$ is effectively redundant, and everywhere it occurs we can simply replace it by $\lambda\underline{b}$.

(ii) Now suppose that $\underline{a}$ and $\underline{b}$ are not parallel, then $\{\underline{a}, \underline{b}\}$ is linearly independent, so that *intuitively* we feel sure that it spans the plane.

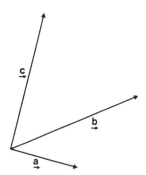

In this case $\underline{c}$ can be expressed in the form $\lambda\underline{a} + \mu\underline{b}$, so that

$$\underline{c} = \lambda\underline{a} + \mu\underline{b},$$

and everywhere $\underline{c}$ occurs it can be replaced by this linear combination of $\underline{a}$ and $\underline{b}$. Here we have made $\underline{c}$ the redundant element.

Notice that in the first case we have $\underline{a} = \lambda\underline{b}$, so that

$$\underline{a} - \lambda\underline{b} + 0\underline{c} = \underline{0},$$

and in the second case $\underline{c} = \lambda\underline{a} + \mu\underline{b}$, so that

$$\underline{c} - \lambda\underline{a} - \mu\underline{b} = \underline{0}$$

From the definition of linear dependence we can see that in both cases the set of geometric vectors $\{a, b, c\}$ is linearly dependent.

It appears that *any* set of three geometric vectors lying in a plane *must* be linearly dependent, and that a linearly independent set of just two geometric vectors is needed to span the set of (plane) geometric vectors. The above argument does not constitute a proof, but our intuitive discussion suggests the result which we shall prove later.

Exercise 1

(i) Are a and $-a$ linearly dependent or independent?
(ii) Are a and 0 linearly dependent or independent? ▪

Exercise 2

If a and b are linearly dependent, are

(i) $3a$ and $4b$ linearly dependent?
(ii) $a + b$ and $a - b$ linearly dependent? ▪

Exercise 3

If a and b are linearly independent, are

(i) $3a$ and $4b$ linearly independent?
(ii) $a + b$ and $a - b$ linearly independent? ▪

Base Geometric Vectors

We shall see later that *any* linearly independent set of two geometric vectors does in fact span the set of (plane) geometric vectors; this leads us naturally to the following definition.

If a subset of a set of geometric vectors spans the whole set, and if in addition the subset is linearly independent, then we say that the subset forms a basis for the set. The elements of the subset are called base geometric vectors, or more briefly base vectors.

It seems clear that there will be two geometric vectors in a basis for the set of plane geometric vectors, and very likely that there will be three geometric vectors in a basis for the set of geometric vectors in three dimensions. We haven't *proved* either of these statements, and there are several vital things to be cleared up. For example, how do we know that *every* basis for a particular set of geometric vectors contains the same number of elements? We shall return to these questions in section 22.2.3.

Exercise 1
(2 minutes)

Exercise 2
(3 minutes)

Exercise 3
(2 minutes)

Definition 6
* * *

Solution 1

(i) $a + (-a) = 0$, and hence a and $-a$ are linearly dependent.

(ii) Choose the values $\alpha = 0$ and $\beta = 1$, say, then

$$\alpha a + \beta 0 = 0a + 0 = 0$$

It follows that a and 0 are linearly dependent. ■

Solution 2

If a and b are linearly dependent, then we know that there are numbers α and β, not both zero, such that

$$\alpha a + \beta b = 0$$

(i) It follows that

$$\frac{\alpha}{3}(3a) + \frac{\beta}{4}(4b) = 0$$

and hence $3a$ and $4b$ are also linearly dependent.

(ii) $a + b$ and $a - b$ are linearly dependent if we can find numbers λ and μ not both zero, such that

$$\lambda(a + b) + \mu(a - b) = 0$$

i.e., such that

$$(\lambda + \mu)a + (\lambda - \mu)b = 0$$

Now a and b are linearly dependent so there are numbers α and β, not both zero, such that

$$\alpha a + \beta b = 0$$

Suppose that we choose $\lambda + \mu = \alpha$
and $\lambda - \mu = \beta$

so that $\lambda = \dfrac{\alpha + \beta}{2}$ and $\mu = \dfrac{\alpha - \beta}{2}$,

then if λ and μ are not both zero we have shown that $a + b$ and $a - b$ are linearly dependent. But this follows at once, since $\lambda = \mu = 0$ implies $\alpha = \beta = 0$, which we know to be false. ■

Solution 3

(i) Suppose that we can find numbers α and β such that

$$\alpha(3a) + \beta(4b) = 0$$

i.e.

$$3\alpha(a) + 4\beta(b) = 0$$

But a and b are linearly independent, so $3\alpha = 4\beta = 0$, which implies that $\alpha = \beta = 0$. So $3a$ and $4b$ are linearly independent.

(ii) Yes. The proof is similar to part (i). ■

22.1.5 An Algebra of Number Pairs

In this section we shall look at the relationship between Cartesian co-ordinates and geometric vectors.

Cartesian Co-ordinates and Geometric Vectors

We choose geometric vectors $\underline{i}$ and $\underline{j}$ of unit length, in the directions of the Cartesian x and y axes respectively.

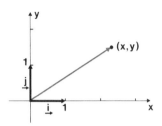

For any point $\underline{q}(0)$ with co-ordinates (x, y) we have

$$\underline{q}(0) = (x\underline{i} + y\underline{j})(0)$$

so that

$$\underline{q} = x\underline{i} + y\underline{j}$$

We have concentrated on geometric vectors, but, since there is a one-one relationship between geometric vectors in a plane and number pairs, we can also construct an *algebra of number pairs.*

Let us start then with the set of all ordered pairs of real numbers, $R \times R$, as the set on which the algebra is to be constructed. We shall call f the mapping from (plane) geometric vectors to rectangular Cartesian co-ordinates (x, y). That is,

$$f : x\underline{i} + y\underline{j} \longmapsto (x, y)$$

We know that f is one-one, so operations on the set of geometric vectors will "carry over" to the set $R \times R$. What happens to addition and multiplication by a scalar under f?

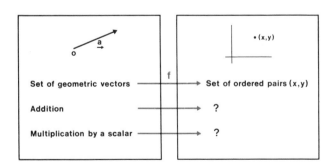

Addition on R × R

Suppose that we are given two geometric vectors $\underline{a}$ and $\underline{b}$ corresponding to the pairs (x_a, y_a) and (x_b, y_b) respectively. Then

$$\underline{a} = x_a\underline{i} + y_a\underline{j}$$

and

$$\underline{b} = x_b\underline{i} + y_b\underline{j}$$

It follows that

$$\underset{\rightarrow}{a} + \underset{\rightarrow}{b} = (x_a\underset{\rightarrow}{i} + y_a\underset{\rightarrow}{j}) + (x_b\underset{\rightarrow}{i} + y_b\underset{\rightarrow}{j})$$
$$= (x_a + x_b)\underset{\rightarrow}{i} + (y_a + y_b)\underset{\rightarrow}{j},$$

using the associativity and commutativity of addition on the set of geometric vectors. The sum $\underset{\rightarrow}{a} + \underset{\rightarrow}{b}$ therefore corresponds to the pair $(x_a + x_b, y_a + y_b)$ and we are led to define the induced addition on $R \times R$ by*

$$(x_a, y_a) + (x_b, y_b) = (x_a + x_b, y_a + y_b)$$

Definition 1
* * *

We have, as it were, used the black arrows in the following commutative diagram to define the $+$.

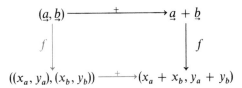

Notice that the two $+$'s are different, being defined on different sets, but because of the way $+$ has been defined as the induced binary operation on $R \times R$, it has the same properties as $+$ on the set of geometric vectors. In particular, it is associative and commutative.

Multiplication by a scalar on $R \times R$

Suppose that the geometric vector $\underset{\rightarrow}{a}$ corresponds to the pair (x_a, y_a); then

$$\underset{\rightarrow}{a} = x_a\underset{\rightarrow}{i} + y_a\underset{\rightarrow}{j},$$

so that

$$\lambda\underset{\rightarrow}{a} = \lambda(x_a\underset{\rightarrow}{i} + y_a\underset{\rightarrow}{j})$$
$$= \lambda x_a\underset{\rightarrow}{i} + \lambda y_a\underset{\rightarrow}{j} \quad \text{(by property (xii) on page 17)}$$

The geometric vector $\lambda\underset{\rightarrow}{a}$ therefore corresponds to the ordered pair $(\lambda x_a, \lambda y_a)$, and we are led to define multiplication by a scalar on $R \times R$ by

Definition 2
* * *

$$\lambda(x_a, y_a) = (\lambda x_a, \lambda y_a)$$

We have the following commutative diagram.

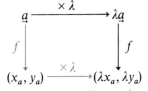

We can, of course, define subtraction by analogy, also retaining all the properties that subtraction has for the set of geometric vectors. We are now able to leave the algebra of geometric vectors if we wish, and concentrate on the algebra of number pairs. But why restrict ourselves to number ·pairs? Why not triples, or n-tuples for that matter? In fact, we need not even restrict ourselves to any of these. Why not discuss some *abstract system* which has the essential properties of the above system? For then any results which we establish *apply to any such system.*

---

* Equality of ordered pairs is of course defined by:

$$(x, y) = (x', y') \Leftrightarrow x = x' \quad \text{and} \quad y = y'.$$

Before we leave the algebra of geometric vectors and plunge into the abstract concept of a *vector space*, we shall look briefly at two other possible lines of advance, each arising from the problem of defining "multiplication" on the set of geometric vectors.

22.1.6 "Multiplication" on the Set of Geometric Vectors

It will be well worth your while to read section 22.1.6 and the Appendix in order to obtain a deeper understanding, but if you are short of time go on to section 22.2.

We shall work in three-dimensional space, because it is here that the concepts are really useful.

In our construction of an algebra on the set of geometric vectors we have so far avoided the problem of defining a binary operation which could correspond to the operation of multiplication on the set of real numbers. It is true that we have defined multiplication by a scalar, but this does not combine two elements from the set of geometric vectors, but simply a geometric vector with a scalar.

There are in fact several different operations which we can define, all having some features in common with multiplication of real numbers. The criteria for any particular choice of operation can be varied, but one criterion high on the list must be usefulness, either within mathematics or in the application of mathematics. We shall choose a particular operation, and we shall demonstrate one of its uses later. (See also Exercise 1.) The chosen operation is called the *inner product*, sometimes called the *scalar product* or *dot product*.

The Inner Product

Suppose that we are given two geometric vectors $\underset{\sim}{a}$ and $\underset{\sim}{b}$ with an angle θ between them (i.e. θ is the angle between any two arrows $\vec{a}$ and $\vec{b}$, where $\vec{a} \in \underset{\sim}{a}$ and $\vec{b} \in \underset{\sim}{b}$).

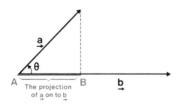

Notice that the angle θ lies between the two arrow heads, *not* like the angle in the following diagram; also, we take θ such that $0 \leqslant \theta \leqslant \pi$.

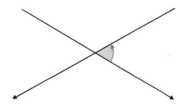

The inner product of $\underline{a}$ and $\underline{b}$, denoted by $\underline{a} \cdot \underline{b}$ is defined by

$$\underline{a} \cdot \underline{b} = |\underline{a}| \, |\underline{b}| \cos \theta$$

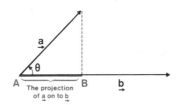

The length AB is called the projection of $\underline{a}$ on to $\underline{b}$ and is equal to $|\underline{a}| \cos \theta$. So the inner product of $\underline{a}$ and $\underline{b}$ is equal to $|\underline{b}|$ times the projection of $\underline{a}$ on to $\underline{b}$, and it is also equal to $|\underline{a}|$ times the projection of $\underline{b}$ on to $\underline{a}$. Notice that this operation is a *commutative binary operation* (just like multiplication of real numbers) but it is *not closed* on the set of geometric vectors, because the combination of two geometric vectors is *not* a geometric vector but a scalar (a real number). This means, among other things, that we cannot define an "inverse" operation adequately, i.e. that there is no possibility of finding an equivalent of division. If we consider what we mean by division on the set of real numbers, we see that it is an operation which "undoes" the work of multiplication. Thus, if we take a real number a and multiply it by a non-zero number b, then divide the result by b, we are back to a. But if we take a geometric vector $\underline{a}$ and form its inner product with a geometric vector $\underline{b}$, then we get $\underline{a} \cdot \underline{b}$ which is a real number, and the problem of getting back from this real number to the geometric vector $\underline{a}$ is not analogous to the multiplication/division problem.

This operation is very useful however. You should notice particularly that:

(i) if θ is the angle between two non-zero geometric vectors $\underline{a}$ and $\underline{b}$, then

$$\cos \theta = \frac{\underline{a} \cdot \underline{b}}{|\underline{a}| \, |\underline{b}|} \; ;$$

(ii) $(\underline{a} \cdot \underline{b} = 0) \Leftrightarrow (\underline{a} \text{ and } \underline{b} \text{ are perpendicular})$
i.e. $\underline{a} \cdot \underline{b} = 0 \Leftrightarrow \cos \theta = 0$
provided, of course, that neither $\underline{a}$ nor $\underline{b}$ is the zero geometric vector;

(iii) $\underline{a} \cdot \underline{a} = |\underline{a}|^2$, since $\cos 0 = 1$.

So if we can find a convenient way of calculating the inner product (as we can) which does not directly involve the definition given above, then we can use the inner product:

  (i) to calculate the angle between two geometric vectors;
 (ii) to determine whether or not two geometric vectors are perpendicular;
(iii) to calculate the length of a geometric vector.

What properties of ordinary multiplication of real numbers does the inner product have?

It follows from the definition of inner product that

    $\underline{a} \cdot \underline{b} = \underline{b} \cdot \underline{a}$, i.e. the inner product is commutative
    $\lambda \underline{a} \cdot \mu \underline{b} = \lambda \mu \underline{a} \cdot \underline{b}$.

Associativity does not apply, since the inner product is not closed.*

Is the inner product distributive over addition of geometric vectors?

* We cannot form $(\underline{a} \cdot \underline{b}) \cdot \underline{c}$ because $(\underline{a} \cdot \underline{b})$ is a *scalar*, not a geometric vector, and we therefore cannot form its inner product with $\underline{c}$, i.e. $(\underline{a} \cdot \underline{b}) \cdot \underline{c}$ is *meaningless*.

The Distributive Laws

Consider the three arbitrary geometric vectors (in three dimensions) $\underset{\rightarrow}{a}$, $\underset{\rightarrow}{b}$ and $\underset{\rightarrow}{c}$ illustrated in the following diagram:

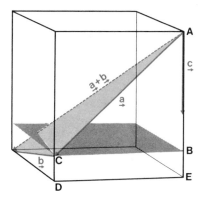

From the diagram we can see that

$$\underrightarrow{AE} = \underrightarrow{AB} + \underrightarrow{CD},$$

so that

(the projection of $(\underset{\rightarrow}{a} + \underset{\rightarrow}{b})$ on to $\underset{\rightarrow}{c}$) =
(the projection of $\underset{\rightarrow}{a}$ on to $\underset{\rightarrow}{c}$) + (the projection of $\underset{\rightarrow}{b}$ on to $\underset{\rightarrow}{c}$).

Multiplying both sides of the equation by $|\underset{\rightarrow}{c}|$, we get

$$(\underset{\rightarrow}{a} + \underset{\rightarrow}{b}) \cdot \underset{\rightarrow}{c} = \underset{\rightarrow}{a} \cdot \underset{\rightarrow}{c} + \underset{\rightarrow}{b} \cdot \underset{\rightarrow}{c}$$

i.e. the inner product is right-distributive over addition of geometric vectors.

Exercise 1

Exercise 1
(2 minutes)

Use the distributive result above and the commutative property of the inner product to show that

$$\underset{\rightarrow}{c} \cdot (\underset{\rightarrow}{a} + \underset{\rightarrow}{b}) = \underset{\rightarrow}{c} \cdot \underset{\rightarrow}{a} + \underset{\rightarrow}{c} \cdot \underset{\rightarrow}{b},$$

i.e. that · is left-distributive over +. ■

The result of Exercise 1 proves that the inner product is distributive over addition of geometric vectors.

Main Text
* * *

We have seen that the inner product has some properties in common with ordinary multiplication, but there are important differences, essentially because the inner product is not closed. Although we have used a dot for the inner product, and we often use a dot for multiplication of real numbers, we must be careful not to make unwarranted deductions. For example, the statement

Discussion

$$(\underset{\rightarrow}{a} \cdot \underset{\rightarrow}{b} = 0) \Rightarrow (\underset{\rightarrow}{a} = \underset{\rightarrow}{0}) \vee (\underset{\rightarrow}{b} = \underset{\rightarrow}{0})$$

is FALSE, because $\underset{\rightarrow}{a} \cdot \underset{\rightarrow}{b} = 0$ also if neither $\underset{\rightarrow}{a}$ nor $\underset{\rightarrow}{b}$ is $\underset{\rightarrow}{0}$ but $\underset{\rightarrow}{a}$ is perpendicular to $\underset{\rightarrow}{b}$. Again the statement

$$(\underset{\rightarrow}{a} \cdot \underset{\rightarrow}{b} = \underset{\rightarrow}{a} \cdot \underset{\rightarrow}{c}) \Rightarrow (\underset{\rightarrow}{b} = \underset{\rightarrow}{c})$$

is also FALSE, because

$$|\underset{\rightarrow}{a}||\underset{\rightarrow}{b}|\cos \theta_1 = |\underset{\rightarrow}{a}||\underset{\rightarrow}{c}|\cos \theta_2,$$

where θ_1, θ_2 are the angles between $\underset{\rightarrow}{a}$ and $\underset{\rightarrow}{b}$, $\underset{\rightarrow}{a}$ and $\underset{\rightarrow}{c}$ respectively, does not imply that

$$\underset{\rightarrow}{b} = \underset{\rightarrow}{c}.$$

(*continued on page 30*)

Solution 1

We have

$$\underset{\sim}{c} \cdot (\underset{\sim}{a} + \underset{\sim}{b}) = (\underset{\sim}{a} + \underset{\sim}{b}) \cdot \underset{\sim}{c} \qquad (\cdot \text{ is commutative})$$
$$= \underset{\sim}{a} \cdot \underset{\sim}{c} + \underset{\sim}{b} \cdot \underset{\sim}{c} \qquad (\cdot \text{ is right-distributive over } +)$$
$$= \underset{\sim}{c} \cdot \underset{\sim}{a} + \underset{\sim}{c} \cdot \underset{\sim}{b} \qquad (\cdot \text{ is commutative})$$

∎

(*continued from page 29*)

The inner product is particularly easy to deal with because of the simple form it takes when the geometric vectors are written in terms of the basis $\{\underset{\sim}{i}, \underset{\sim}{j}, \underset{\sim}{k}\}$ consisting of geometric vectors of unit length in the directions of the x, y and z Cartesian axes.

Main Text
* * *

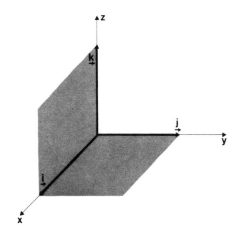

We have:

$$|\underset{\sim}{i}| = |\underset{\sim}{j}| = |\underset{\sim}{k}| = 1;$$

the angle between any base vector and itself is zero;

the angle between any two distinct base vectors is $\dfrac{\pi}{2}$.

From the definition of the inner product we can see that

$$\underset{\sim}{i} \cdot \underset{\sim}{i} = 1 \quad \text{and} \quad \underset{\sim}{i} \cdot \underset{\sim}{j} = \underset{\sim}{j} \cdot \underset{\sim}{i} = 0$$
$$\underset{\sim}{j} \cdot \underset{\sim}{j} = 1 \qquad \underset{\sim}{j} \cdot \underset{\sim}{k} = \underset{\sim}{k} \cdot \underset{\sim}{j} = 0$$
$$\underset{\sim}{k} \cdot \underset{\sim}{k} = 1 \qquad \underset{\sim}{k} \cdot \underset{\sim}{i} = \underset{\sim}{i} \cdot \underset{\sim}{k} = 0.$$

* * *

Suppose now that

$$\underset{\sim}{a} = x_a \underset{\sim}{i} + y_a \underset{\sim}{j} + z_a \underset{\sim}{k}$$

and

$$\underset{\sim}{b} = x_b \underset{\sim}{i} + y_b \underset{\sim}{j} + z_b \underset{\sim}{k}.$$

Using the above results for the inner products involving $\underset{\sim}{i}$, $\underset{\sim}{j}$ and $\underset{\sim}{k}$, and the distributive law, we obtain

$$(x_a \underset{\sim}{i} + y_a \underset{\sim}{j} + z_a \underset{\sim}{k}) \cdot (x_b \underset{\sim}{i} + y_b \underset{\sim}{j} + z_b \underset{\sim}{k})$$
$$= x_a x_b + y_a y_b + z_a z_b,$$

so that we obtain the important formula

$$\underline{a} \cdot \underline{b} = x_a x_b + y_a y_b + z_a z_b.$$

Example 1

Example 1

Find the angle between $\underline{a}$ and $\underline{b}$ if

$$\underline{a} = \underline{i} + 2\underline{j} - 3\underline{k}$$

and $\underline{b} = 2\underline{i} - 5\underline{j} + \underline{k}.$

We know that if θ is the angle between $\underline{a}$ and $\underline{b}$ then

$$\cos \theta = \frac{\underline{a} \cdot \underline{b}}{|\underline{a}| |\underline{b}|}.$$

Now

$$\underline{a} \cdot \underline{b} = 1 \times 2 + 2 \times (-5) + (-3) \times 1 = -11$$

and we know that $\underline{a} \cdot \underline{a} = |\underline{a}|^2$, so that

$$|\underline{a}|^2 = 1^2 + 2^2 + (-3)^2 = 14$$
$$|\underline{b}|^2 = 2^2 + (-5)^2 + 1^2 = 30.$$

Hence

$$\cos \theta = \frac{-11}{\sqrt{14}\sqrt{30}}$$

Since $\cos \theta$ is negative, $\frac{\pi}{2} \leqslant \theta \leqslant \pi$, and in fact θ is approximately $132°$.

■

Example 2

Example 2

Find the angle between a diagonal of a cube and one of its edges.

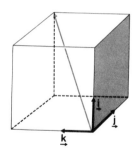

We choose unit vectors $\underline{i}, \underline{j}, \underline{k}$ in the directions of the edges of the cube, then the geometric vector $\underline{a} = \underline{i} + \underline{j} + \underline{k}$ has the same direction as a diagonal of the cube, and

$$\underline{a} \cdot \underline{i} = |\underline{a}| \cos \theta$$

where θ is the acute angle we wish to determine.

$$\underline{a} \cdot \underline{i} = (\underline{i} + \underline{j} + \underline{k}) \cdot \underline{i} = 1$$

also

$$|\underline{a}|^2 = \underline{a} \cdot \underline{a} = 3$$

so that $\sqrt{3} \cos \theta = 1$, i.e. $\cos \theta = \dfrac{1}{\sqrt{3}}$, so that θ is approximately $55°$.

■

22.2   VECTOR SPACES

22.2.1   The Algebra of Lists

On various occasions in the previous sections we saw that if we have a co-ordinate system, then there is a one-one correspondence between geometric vectors in two (or three) dimensions and ordered pairs (or triples) of numbers.

We developed some ways of combining geometric vectors, that is to say, we developed an algebra of geometric vectors, and we showed that there was a corresponding algebra for the ordered pairs or triples. For example, corresponding to the result

$$(a_1\underline{i} + a_2\underline{j} + a_3\underline{k}) + (b_1\underline{i} + b_2\underline{j} + b_3\underline{k})$$
$$= (a_1 + b_1)\underline{i} + (a_2 + b_2)\underline{j} + (a_3 + b_3)\underline{k}$$

we have

$$(a_1, a_2, a_3) + (b_1, b_2, b_3) = (a_1 + b_1, a_2 + b_2, a_3 + b_3).$$

We can regard these triples just as *ordered lists* and write them in any convenient form; for example, we can write

$$\begin{pmatrix} a_1 \\ a_2 \\ a_3 \end{pmatrix} + \begin{pmatrix} b_1 \\ b_2 \\ b_3 \end{pmatrix} = \begin{pmatrix} a_1 + b_1 \\ a_2 + b_2 \\ a_3 + b_3 \end{pmatrix}.$$

Equation (1)

Such "vertical" lists are used later when we discuss matrices.

In the same way, the multiplication of a geometric vector by a scalar corresponds to

$$\lambda \begin{pmatrix} a_1 \\ a_2 \\ a_3 \end{pmatrix} = \begin{pmatrix} \lambda a_1 \\ \lambda a_2 \\ \lambda a_3 \end{pmatrix}.$$

Equation (2)

So far these lists are just a way of specifying geometric vectors, but do they only give us an alternative notation, or do they suggest anything new? Let's forget for a moment the origins of the lists. Equations (1) and (2) define ways of manipulating lists of numbers. There is no reason why we should always have only two or three elements in the list. Equations (1) and (2) can be extended to lists with more than three elements; for example, we can write

$$\begin{pmatrix} a_1 \\ a_2 \\ a_3 \\ \vdots \\ a_n \end{pmatrix} + \begin{pmatrix} b_1 \\ b_2 \\ b_3 \\ \vdots \\ b_n \end{pmatrix} = \begin{pmatrix} a_1 + b_1 \\ a_2 + b_2 \\ a_3 + b_3 \\ \vdots \\ a_n + b_n \end{pmatrix}$$

and

$$\lambda \begin{pmatrix} a_1 \\ a_2 \\ a_3 \\ \vdots \\ a_n \end{pmatrix} = \begin{pmatrix} \lambda a_1 \\ \lambda a_2 \\ \lambda a_3 \\ \vdots \\ \lambda a_n \end{pmatrix}.$$

But this is rather futile if we only have a physical or mathematical interpretation when the lists contain not more than three elements. However, we can use these lists to describe situations other than the algebra of geometric vectors, and we can interpret results and concepts in one situation (for example, basis and linear independence) to give results and concepts in another. That is, we can establish *morphism* from one structure to another, and we shall go on to discuss the abstract structure which typifies all the exemplary situations.

What else can we represent by lists?

Example 1 Polynomial Functions   Example 1

Consider the set of all polynomial functions of the form

$$p : x \longmapsto ax^3 + bx^2 + cx + d \qquad (x \in R)$$

where a, b, c and d are real numbers.

We can represent p by the four coefficients a, b, c and d, which we can arrange as a list:

$$\begin{pmatrix} a \\ b \\ c \\ d \end{pmatrix}.$$

The addition of two such polynomial functions corresponds to the addition of the corresponding two lists. Thus if

$$p_1 : x \longmapsto a_1 x^3 + b_1 x^2 + c_1 x + d_1 \qquad (x \in R)$$

and

$$p_2 : x \longmapsto a_2 x^3 + b_2 x^2 + c_2 x + d_2 \qquad (x \in R)$$

then

$p_1 + p_2$ corresponds to the list

$$\begin{pmatrix} a_1 \\ b_1 \\ c_1 \\ d_1 \end{pmatrix} + \begin{pmatrix} a_2 \\ b_2 \\ c_2 \\ d_2 \end{pmatrix} = \begin{pmatrix} a_1 + b_1 \\ a_2 + b_2 \\ c_1 + c_2 \\ d_1 + d_2 \end{pmatrix},$$

and the function λp corresponds to the list

$$\lambda \begin{pmatrix} a \\ b \\ c \\ d \end{pmatrix} = \begin{pmatrix} \lambda a \\ \lambda b \\ \lambda c \\ \lambda d \end{pmatrix}.$$

(Remember that a, b, c and d can be *any* real numbers, and so a function such as $f : x \longmapsto 0x^3 + 0x^2 + x + 1$ is included in this set of functions.) By considering polynomials of degree higher than three we would get examples of lists with more than four elements. ∎

Example 2 Finite Sequences   Example 2

Consider the set of all finite sequences of real numbers with, say, n terms.

The methods of adding finite sequences and multiplying them by a number, which we met in *Unit 7, Sequences and Limits I*, can be seen to correspond to our algebra of lists, simply by rewriting a sequence

$$u_1, u_2, u_3, \cdots, u_n$$

in the form of a list

$$\begin{pmatrix} u_1 \\ u_2 \\ u_3 \\ \vdots \\ u_n \end{pmatrix}.$$

■

Example 3   Shopping Lists

Example 3

Ordinary shopping lists can be combined by these algebraic operations. For example, we may be in the habit of ordering from the milkman: silver top milk, gold top milk, red top milk and cartons of cream. Then if the list

$$\begin{pmatrix} 3 \\ 1 \\ 2 \\ 0 \end{pmatrix}$$

represents an order of 3 silver top, 1 gold top, 2 red top and no cream, a week's supplies may be added together like this:

$$\begin{pmatrix} 2 \\ 2 \\ 1 \\ 1 \end{pmatrix} + \begin{pmatrix} 1 \\ 2 \\ 0 \\ 0 \end{pmatrix} + \begin{pmatrix} 1 \\ 2 \\ 0 \\ 1 \end{pmatrix} + \begin{pmatrix} 1 \\ 2 \\ 0 \\ 1 \end{pmatrix} + \begin{pmatrix} 1 \\ 2 \\ 1 \\ 0 \end{pmatrix} + \begin{pmatrix} 2 \\ 2 \\ 0 \\ 0 \end{pmatrix} + \begin{pmatrix} 1 \\ 2 \\ 1 \\ 1 \end{pmatrix} = \begin{pmatrix} 9 \\ 14 \\ 3 \\ 4 \end{pmatrix}.$$

This tells us that we order 9 silver top, 14 gold top, 3 red top and 4 cartons of cream. Of course, if we have the same order every day, the calculation can be speeded up. For example, we may have

$$7 \begin{pmatrix} 2 \\ 1 \\ 0 \\ 0 \end{pmatrix} = \begin{pmatrix} 14 \\ 7 \\ 0 \\ 0 \end{pmatrix}.$$

In this case we can see that the following four lists,

$$\begin{pmatrix} 1 \\ 0 \\ 0 \\ 0 \end{pmatrix} \begin{pmatrix} 0 \\ 1 \\ 0 \\ 0 \end{pmatrix} \begin{pmatrix} 0 \\ 0 \\ 1 \\ 0 \end{pmatrix} \begin{pmatrix} 0 \\ 0 \\ 0 \\ 1 \end{pmatrix},$$

representing each of the basic commodities which we order, can be used as "basic building bricks" to specify *any* order to the milkman. This conforms with the idea of a basis for the set of geometric vectors.   ■

Example 4   Solution of Differential Equations

Example 4

In Unit 12, section 12.2.3, we introduced the differentiation operator

$$D: f \longmapsto f',$$

with domain the set of all real functions. We often write $D^2 f = f''$, and similarly for higher derivatives. With this notation we are able to discuss more complicated operators such as

$$D^2 - 3D + 2,$$

which is defined by

$$(D^2 - 3D + 2): f \longmapsto f'' - 3f' + 2f.$$

Often in applied mathematics we are faced with the problem of finding a function which is mapped to a given function, g say, under such an operator. In other words, what function (or functions) f satisfy the equation

$$(D^2 - 3D + 2)f = g?$$

In terms of image values we require that

$$D^2 f(x) - 3Df(x) + 2f(x) = g(x).$$

Equations of this kind are called *differential equations*. Suppose, for example, that g is the zero function, i.e.

$$g: x \longmapsto 0 \qquad (x \in R)$$

then

$$f_1: x \longmapsto e^x \qquad (x \in R)$$

is one function which satisfies this equation.

We know that

$$f_1''(x) = e^x \quad \text{and} \quad f_1'(x) = e^x$$

so that

$$(D^2 - 3D + 2)f(x) = e^x - 3e^x + 2e^x = 0.$$

Another function which satisfies the equation is $f_2: x \longmapsto e^{2x} \qquad (x \in R)$.

We shall show later in this course that any solution of this differential equation has the form

$$\alpha f_1 + \beta f_2$$

where α and β are real numbers, and f_1 and f_2 are the functions given above. If we take f_1 and f_2 as *basic solutions*, then we see that any solution of the form $\alpha f_1 + \beta f_2$ can be represented by the list $\begin{pmatrix} \alpha \\ \beta \end{pmatrix}$. The particular solutions f_1 and f_2 can be represented by $\begin{pmatrix} 1 \\ 0 \end{pmatrix}$ and $\begin{pmatrix} 0 \\ 1 \end{pmatrix}$ respectively, and in general the list $\begin{pmatrix} \alpha \\ \beta \end{pmatrix}$ represents the function

$$x \longmapsto \alpha e^x + \beta e^{2x} \qquad (x \in R).$$

You may like to verify that the lists

$$\begin{pmatrix} \alpha \\ \beta \end{pmatrix} + \begin{pmatrix} \gamma \\ \delta \end{pmatrix} = \begin{pmatrix} \alpha + \gamma \\ \beta + \delta \end{pmatrix}$$

and

$$\lambda \begin{pmatrix} \alpha \\ \beta \end{pmatrix} = \begin{pmatrix} \lambda \alpha \\ \lambda \beta \end{pmatrix}$$

also represent solutions of the equation. ■

Example 5   Magic Squares   **Example 5**

Arrays of numbers such as

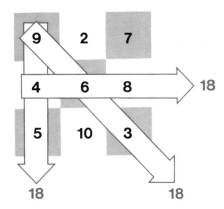

where the rows, columns and diagonals all add up to the same number, are called 3×3 (read as "3 by 3") *magic squares*. If we denote the following magic squares:

1	1	1		0	1	-1		1	-1	0
1	1	1		-1	0	1		-1	0	1
1	1	1		1	-1	0		0	1	-1

by M_1, M_2 and M_3 respectively, then we can represent any magic square with real elements in terms of M_1, M_2 and M_3. For example,

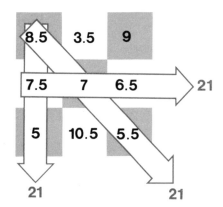

corresponds to $7M_1 - 2M_2 + \frac{3}{2}M_3$ (where, for example, $7M_1$ represents the square:

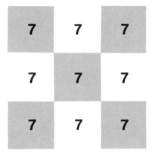

and so on). "Addition" of the squares is performed simply by adding corresponding entries. For example, $M_1 + M_2$ corresponds to the square:

1+0	1+1	1+(-1)		1	2	0
1+(-1)	1+0	1+1	=	0	1	2
1+1	1+(-1)	1+0		2	0	1

and so on. Notice that the result of combining squares by either of these two operations always corresponds to an array of numbers which is itself a magic square.

This example may seem to you very much like a rabbit coming out of a hat: why did we choose these three squares as building blocks? And why three? Why not two, or four? It turns out that the set of 3×3 magic squares forms what we call a *vector space of three dimensions*, and this is why we need *three* elements (just as we need three base vectors for geometric vectors in three dimensions). But it is not worth discussing this particular system in more detail: it is much easier to appreciate *all* these examples when we have studied the abstract system. ■

Example 6   Interpolating Polynomials   **Example 6**

The functions:

$$f_1 : x \longmapsto \tfrac{1}{2}x^2 - \tfrac{1}{2}x \qquad (x \in R)$$
$$f_2 : x \longmapsto -x^2 + 1 \qquad (x \in R)$$
$$f_3 : x \longmapsto \tfrac{1}{2}x^2 + \tfrac{1}{2}x \qquad (x \in R)$$

have an interesting property. If we tabulate the images at -1, 0 and 1, we get the following table:

x	-1	0	1
$f_1(x)$	1	0	0
$f_2(x)$	0	1	0
$f_3(x)$	0	0	1

We can use these three functions to write down, without calculation, a formula for the quadratic function which has given values at $-1, 0$ and 1. For example, the quadratic function f which takes the values 2, 3 and 6 at $-1, 0$ and 1 respectively, is given by

$$f : x \longmapsto 2f_1(x) + 3f_2(x) + 6f_3(x) \qquad (x \in R)$$

i.e.

$$f : x \longmapsto x^2 + 2x + 3 \qquad (x \in R)$$

We can use the idea of this example to derive Simpson's rule (see *Unit 9, Integration I*, section 9.4.2) very quickly. Suppose a curve passes through the three points $(-1, g_{-1}), (0, g_0), (1, g_1)$. Then we know from this example that if we approximate to this curve by a parabola passing through these points, then that parabola is the graph of the function

$$g : x \longmapsto g_{-1} f_1(x) + g_0 f_2(x) + g_1 f_3(x) \qquad (x \in R)$$

i.e.

$$g : x \longmapsto g_{-1}(\tfrac{1}{2}x^2 - \tfrac{1}{2}x) + g_0(-x^2 + 1) + g_1(\tfrac{1}{2}x^2 + \tfrac{1}{2}x),$$

i.e.

$$g : x \longmapsto x^2(\tfrac{1}{2}g_{-1} - g_0 + \tfrac{1}{2}g_1) + x(\tfrac{1}{2}g_1 - \tfrac{1}{2}g_{-1}) + g_0.$$

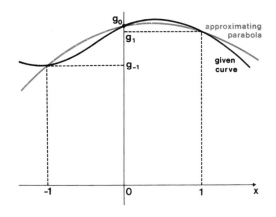

The area bounded by the parabola, the x-axis and the lines defined by $x = -1$ and $x = 1$ is

$$\int_{-1}^{1} g = \left[x \longmapsto \frac{x^3}{3}(\tfrac{1}{2}g_{-1} - g_0 + \tfrac{1}{2}g_1) \right.$$

$$\left. + \frac{x^2}{2}(\tfrac{1}{2}g_1 - \tfrac{1}{2}g_{-1}) + g_0 x \right]_{-1}^{1}$$

$$= \tfrac{1}{3}(g_{-1} + 4g_0 + g_1)$$

which is Simpson's rule with unit sub-interval length. (You may like to see if you can make the very slight modification to this piece of work which gives the more general form of Simpson's rule, where the strips are of width h.) ∎

We shall meet more examples of "lists" in this course, and if you go further in mathematics you will see that this "algebra of lists" is far more important and rich in material than may be apparent at the moment.

When a unifying idea emerges, it is often useful to strip it of all its original trappings and try to express the real essence of the idea. So we shall now extract the essential properties from our discussion so far.

Discussion
* *

22.2.2 Vector Spaces

We now generalize the discussion to an arbitrary set of elements, and construct a very general mathematical structure called a *vector space*.

A feature common to the geometric vectors and all the examples in the last section, is that, in each case, we had a set on which we could sensibly define *addition* and *multiplication by a scalar*.

We shall take the structure which we have developed on the set of geometric vectors as our model, and discuss an arbitrary set with operations called *addition* and *multiplication by a scalar* defined on it. If the structure satisfies the following axioms, then we call it a vector space, and we call its elements vectors.

The set of geometric vectors is a particular example of a vector space, and it is the origin of the subject in geometry which motivates this use of the word *space*.

One of the purposes of talking about structures such as vector spaces in the abstract is that we hope to be able to represent a number of apparently different structures in the same terms. Thus when we refer to a vector in a vector space, it may be a geometric vector, a solution to a differential equation, a magic square, a milkman's order or one of many other things. If we wish to prove theorems about all these situations at once (by proving them about vector spaces in general), then we choose a notation which is not too suggestive of any one example, and yet does not lose entirely the connection with our principal example, in this case geometric vectors. We therefore use underlined, lower case letters such as $\underline{a}$ to represent a vector.

If $\underline{v}, \underline{v}_1, \underline{v}_2, \underline{v}_3$ are any elements of a set V, and α, β are any real numbers, we require the operations of *addition* of elements of V and *multiplication* of elements of V by a scalar to have the following properties:

Axioms of a Vector Space

1 $\underline{v}_1 + \underline{v}_2$ is a unique element of V
(V is closed for addition)
2 $\underline{v}_1 + (\underline{v}_2 + \underline{v}_3) = (\underline{v}_1 + \underline{v}_2) + \underline{v}_3$
(addition is associative)
3 $\underline{v}_1 + \underline{v}_2 = \underline{v}_2 + \underline{v}_1$
(addition is commutative)
4 There is an element in V, which we call $\underline{v}_0$, such that

$$\underline{v} + \underline{v}_0 = \underline{v}$$

5 $\alpha \underline{v}$ is an element of V
6 $\underline{v} + (-1)\underline{v} = \underline{v}_0$
7 $\alpha(\underline{v}_1 + \underline{v}_2) = (\alpha\underline{v}_1) + (\alpha\underline{v}_2)$
8 $(\alpha + \beta)\underline{v} = \alpha\underline{v} + \beta\underline{v}$
9 $(\alpha\beta)\underline{v} = \alpha(\beta\underline{v})$
10 $1 \times \underline{v} = \underline{v}$

These ten axioms are the axioms of a vector space. There are two important points to note. Strictly speaking, we should call V a vector space over the real numbers or a real vector space, because vector spaces exist involving sets of scalars other than the set of real numbers; we shall discuss only vector spaces over the real numbers. Secondly, we have taken as implicit all the relevant properties of the real numbers*, and these should really be stated along with the other axioms. Any other set with these properties can be taken as the set of scalars in place of the set of real numbers to give a different vector space.

* i.e. the properties Re(1)–Re(4) listed in section 6.1.1 of *Unit 6, Inequalities.*

The axioms of a vector space therefore consist of three sets of axioms:

(i) those applying to the set of vectors only
   (1 to 4 above);
(ii) those applying to the set of scalars only
   (not stated above: the missing axioms are the axioms of what is known in mathematics as a *field*);
(iii) those which describe the interaction between the set of scalars (field) and the set of vectors
   (5 to 10 above).

We define an operation of subtraction of vectors by

$$\underline{v}_1 - \underline{v}_2 = \underline{v}_1 + (-1)\underline{v}_2.$$

From axiom 6, it follows that $\underline{v}_1 - \underline{v}_1 = \underline{v}_0$.

The Zero Element

In a vector space, an element $\underline{v}_0$ which satisfies axiom 4 is called a *zero element*. It follows from the axioms that in any vector space V there is only *one* zero element. For suppose there are two vectors $\underline{v}_0$ and $\underline{v}_0'$ which satisfy axiom 4. That is,

$$\underline{v} + \underline{v}_0 = \underline{v}$$

$$\underline{v} + \underline{v}_0' = \underline{v},$$

where, in each equation, $\underline{v}$ is *any* element of V. Let us put $\underline{v} = \underline{v}_0'$ in the first equation, and $\underline{v} = \underline{v}_0$ in the second equation. We obtain

$$\underline{v}_0' + \underline{v}_0 = \underline{v}_0'$$

$$\underline{v}_0 + \underline{v}_0' = \underline{v}_0.$$

By axiom 3,

$$\underline{v}_0' + \underline{v}_0 = \underline{v}_0 + \underline{v}_0'$$

i.e.

$$\underline{v}_0 = \underline{v}_0',$$

so the zero element is *unique*.

Since the zero element in a vector space behaves just like the zero geometric vector, we shall call this element the zero vector and denote it by $\underline{0}$, just as we had $\underline{0}$ for the zero geometric vector. (Remember that, in terms of lists, $\underline{0}$ is the list in which every entry is zero.)

Notation 2
* * *

Further properties of $\underline{0}$ can be deduced from the axioms. For example, putting $\underline{v}_2 = \underline{0}$ in axiom 7, we obtain

$$\alpha(\underline{v}_1 + \underline{0}) = (\alpha\underline{v}_1) + (\alpha\underline{0}).$$

By axiom 4,

$$\alpha\underline{v}_1 = \alpha\underline{v}_1 + \alpha\underline{0}$$

Now we add $(-1)\alpha\underline{v}_1$ to both sides, and use axioms 2 and 3 to give

$$\alpha\underline{v}_1 + (-1)\alpha\underline{v}_1 = (\alpha\underline{v}_1 + (-1)\alpha\underline{v}_1) + \alpha\underline{0}$$

By axiom 6, we have

$$\underline{0} = \underline{0} + \alpha\underline{0}$$

Using axioms 3 and 4 and interchanging the sides of the equation, gives

$$\alpha\underline{0} = \underline{0},$$

where α is any real number.

Exercise 1

 (i) Which of the examples of section 22.2.1 describe vector spaces?

(ii) The set of all polynomial functions of degree n with the operations of addition of functions and multiplication of a function by a real number is not a vector space. Why not? Suggest a suitable modification to make it a vector space.

(HINT: A (real) polynomial function of degree n is a function of the form

$$x \longmapsto a_n x^n + a_{n-1} x^{n-1} + \cdots + a_1 x + a_0 \qquad (x \in R)$$

in which the a_i are real numbers $(i = 1, 2, \ldots, n)$ and $a_n \neq 0$.) ■

Exercise 2

In each of the following cases state whether the given set of lists forms a vector space for the operations of addition of lists and multiplication of a list by a scalar. In each case give reasons for your answer.

 (i) The set of all lists $\begin{pmatrix} x_1 \\ x_2 \\ x_3 \end{pmatrix}$, where x_1, x_2 are x_3 are positive real numbers.

(ii) The set of all lists $\begin{pmatrix} x_1 \\ x_2 \end{pmatrix}$, where x_1, x_2 are real numbers and $x_1 + x_2 = 0$.

(iii) The set of all lists $\begin{pmatrix} x_1 \\ x_2 \end{pmatrix}$, where x_1 and x_2 are real numbers and $x_1 < x_2$.

(iv) The set of all lists $\begin{pmatrix} x_1 \\ x_2 \\ x_3 \end{pmatrix}$, where x_1, x_2 and x_3 are real numbers such that the function

$$f : t \longmapsto x_1 t^2 + x_2 t + x_3 \qquad (t \in R)$$

satisfies $f(k) = 0$, where k is a fixed real number. ■

Exercise 3

If V is a vector space with zero vector $\underline{0}$, show that $\{\underline{0}\}$ is also a vector space. ■

Solution 1
Solution 1

(i) Except for Example 3 (the milkman example), all the examples describe vector spaces. In the case of the milkman's order, there is no reasonable interpretation for, say, $0.07 \begin{pmatrix} 1 \\ 0 \\ 0 \\ 0 \end{pmatrix}$, so the fifth axiom of a vector space is not satisfied for all real α.

(Example 1 is discussed by implication below.)

(ii) The problem is caused when we add, say, the polynomial function $x \longmapsto -x^n$ to the polynomial function $x \longmapsto x^n + x^{n-1}$. Both are of degree n, but their sum is the polynomial $x \longmapsto x^{n-1}$, which is of degree $n - 1$, so $+$ is not closed, i.e. axiom 1 is violated. A simple modification is to consider the set of polynomials of degree *less than or equal to n*. With the suggested operations, this set is indeed a vector space. ∎

Solution 2
Solution 2

(i) No. For example, multiplication by a negative scalar takes us out of the set, i.e. axiom 5 is violated.

(ii) Yes. All the axioms are satisfied. (In fact, all the points corresponding to the vectors lie on the line defined by the equation $y + x = 0$.)

(iii) No. For example, if $x_1 < x_2$ and $\alpha < 0$, then $\alpha x_1 > \alpha x_2$, i.e. $\alpha \begin{pmatrix} x_1 \\ x_2 \end{pmatrix}$ does not belong to the given set, so axiom 5 is violated.

(iv) Yes. All the axioms of a vector space are satisfied. (Each function has a graph which passes through the fixed point $(k, 0)$.) ∎

Solution 3
Solution 3

We check that the axioms of a vector space are satisfied.
1 $\underline{0} + \underline{0} = \underline{0}$, since $\underline{0}$ is the zero element of V, so $\{\underline{0}\}$ is closed for addition.
4 $\underline{0} \in \{\underline{0}\}$.
5 $\alpha \underline{0} = \underline{0} \in \{\underline{0}\}$ (proved in text).

The other axioms are automatically satisfied, since they are satisfied for all elements of V, and $\underline{0} \in V$,

Hence $\{\underline{0}\}$ is a (real) vector space.

Where Next?

In the case of geometric vectors, we introduced the idea of a *basis*. The development of this idea depended on the concepts of linear combination of vectors and linear dependence. We can extend these ideas to the more general concept of a vector space.

We also made passing reference to these ideas in our examples in section 22.2.1. In the differential equation example, we shall see that every solution of

$$D^2f(x) - 3Df(x) + 2f(x) = 0$$

can be represented in terms of two basic solutions, for example

$$f_1 : x \longmapsto e^x \qquad (x \in R)$$
$$f_2 : x \longmapsto e^{2x} \qquad (x \in R).$$

In the magic squares example, every 3 by 3 magic square can be represented in terms of three basic squares: for example,

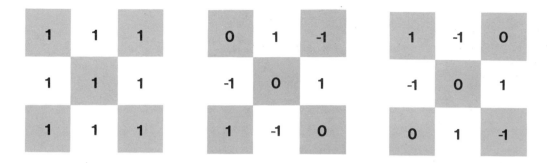

These examples raise a number of questions. How can we choose elements to use as a basis? How many elements do we need? If we can settle the question of how many elements we need — can we select that number of elements at random? Is there any test to see whether or not an arbitrarily chosen set of elements of the right number will do the job?

Before we extend our idea of a *basis* to an abstract vector space, we shall define linear dependence and independence in this context.

Linear Dependence and Independence

The following definitions generalize the notion of linear dependence which we introduced for geometric vectors.

If $v_1, v_2, v_3, \ldots, v_n$ are vectors from any vector space, then an expression of the form

$$\alpha_1 v_1 + \alpha_2 v_2 + \alpha_3 v_3 + \cdots + \alpha_n v_n,$$

where the α's are real numbers, is called a linear combination of vectors.

Definition 2
* * *

The set of vectors $\{v_1, v_2, \ldots, v_n\}$ is said to be linearly dependent if and only if there exist real numbers $\alpha_1, \alpha_2, \ldots, \alpha_n$, which are not all zero, such that

Definition 3
* * *

$$\alpha_1 v_1 + \alpha_2 v_2 + \alpha_3 v_3 + \cdots + \alpha_n v_n = \underline{0}.$$

A set of vectors which is not linearly dependent is said to be *linearly independent*. We can define this term in a more positive way as follows.

A set of vectors $\{v_1, v_2, v_3, \ldots, v_n\}$ is linearly independent if and only if

Definition 4
* * *

$$\alpha_1 v_1 + \alpha_2 v_2 + \alpha_3 v_3 + \cdots + \alpha_n v_n = \underline{0}$$

$$\Rightarrow \alpha_1 = \alpha_2 = \alpha_3 = \cdots = \alpha_n = 0.$$

Remember that we use the terms *dependent* and *independent* in this way because we can express some members of a linearly dependent set in terms of the others. For example, if α_1 is not zero, we can use the axioms of a vector space to write

$$\alpha_1 v_1 + \alpha_2 v_2 + \alpha_3 v_3 + \cdots + \alpha_n v_n = \underline{0}$$

in the form

$$\alpha_1 v_1 = (-\alpha_2) v_2 + (-\alpha_3) v_3 + \cdots + (-\alpha_n) v_n,$$

and then divide by α_1 to give:

$$v_1 = \frac{-\alpha_2}{\alpha_1} v_2 + \frac{-\alpha_3}{\alpha_1} v_3 + \cdots + \frac{-\alpha_n}{\alpha_1} v_n,$$

i.e. v_1 depends on the other vectors. In general, if a set of vectors is linearly dependent, *some* of the vectors in the set (not necessarily every vector, because *some* of the α's may be zero) can be expressed in terms of the others. In other words, some of the elements in the set are redundant.

Exercise 4

Exercise 4
(4 minutes)

In each of the following parts a set of vectors is given. In each case state whether or not the set is linearly independent. In those cases where the set is linearly dependent, express one of the vectors in the set as a linear combination of the others.

(i)
$$\begin{pmatrix} 1 \\ -1 \\ 0 \end{pmatrix}, \quad \begin{pmatrix} 0 \\ 1 \\ 0 \end{pmatrix}, \quad \begin{pmatrix} 1 \\ 0 \\ 0 \end{pmatrix},$$

with the usual operations of addition and multiplication by a scalar for lists. The zero vector in this case is $\begin{pmatrix} 0 \\ 0 \\ 0 \end{pmatrix}$.

(ii) The functions

$$f : x \longmapsto x \qquad (x \in R),$$
$$g : x \longmapsto x^2 \qquad (x \in R),$$

with the operations of addition of functions and multiplication of a function by a real number. The zero vector in this case is

$$\underline{0} : x \longmapsto 0 \qquad (x \in R).$$

(iii)
$$\begin{pmatrix} 2 \\ 0 \\ 0 \end{pmatrix}, \quad \begin{pmatrix} 0 \\ 3 \\ 0 \end{pmatrix}, \quad \begin{pmatrix} 0 \\ 0 \\ 5 \end{pmatrix},$$

with the usual operations for combining "lists". ∎

Exercise 5

Exercise 5
(3 minutes)

If the set of vectors $\underline{v}_1, \underline{v}_2, \underline{v}_3, \dots, \underline{v}_n$ is linearly independent, show that if

$$\alpha_1 \underline{v}_1 + \alpha_2 \underline{v}_2 + \alpha_3 \underline{v}_3 + \cdots + \alpha_n \underline{v}_n = \beta_1 \underline{v}_1 + \beta_2 \underline{v}_2 + \cdots + \beta_n \underline{v}_n$$

then

$$\alpha_1 = \beta_1, \qquad \alpha_2 = \beta_2, \dots, \alpha_n = \beta_n.$$

Notice that this result implies that a vector $\underline{v}$ cannot be expressed in two *different* ways as a linear combination of a set of linearly independent vectors. ∎

Exercise 6

Exercise 6
(5 minutes)

If $\{\underline{v}_1, \underline{v}_2, \dots, \underline{v}_n\}$ is a linearly independent set of vectors, prove that any subset of this set is also linearly independent. ∎

Exercise 7

Exercise 7
(5 minutes)

If $\{\underline{v}_1, \underline{v}_2, \dots, \underline{v}_n\}$ is a linearly dependent subset of a vector space V, prove that $\{\underline{v}_1, \underline{v}_2, \dots, \underline{v}_n, \underline{w}\}$ is also linearly dependent, where $\underline{w}$ is *any* element in V. ∎

22.2.3 Bases and Dimension of a Vector Space

In section 22.1.4 we saw that it is possible to select two geometric vectors in a plane, and then to specify every geometric vector in the plane as a linear combination of those two. Similarly, in three dimensions we need to select three geometric vectors. We called such a set a *basis*, and we now wish to extend the same idea to a vector space.

The set of vectors $\{v_1, v_2, \ldots, v_m\}$ is said to span the vector space V if for each element $\underline{w}$ in V we can find scalars $\alpha_1, \alpha_2, \ldots, \alpha_m$, such that

$$\underline{w} = \alpha_1 v_1 + \alpha_2 v_2 + \alpha_3 v_3 + \cdots + \alpha_m v_m.$$

If the set of vectors $\{v_1, v_2, \ldots, v_n\}$ is linearly independent and spans the vector space V, then we say that it forms a basis for V, and $v_1, v_2, \ldots, v_n$ are called base vectors.

Essentially a basis contains the minimum number of elements which are required to span the space. In exercise 22.2.2.5 we saw that any vector can be expressed in a *unique* way as a linear combination of the elements of a basis.

For example, the set $\{\underline{i}, \underline{j}, \underline{k}\}$ spans the three-dimensional geometric vector space, because each geometric vector $\underline{r}$ can be expressed in the form

$$\underline{r} = x\underline{i} + y\underline{j} + z\underline{k}.$$

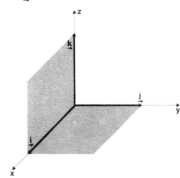

Here $\underline{i}, \underline{j}$ and $\underline{k}$ play the parts of v_1, v_2 and v_3, and we know that it is possible to find the appropriate values x, y and z which play the parts of α_1, α_2 and α_3. Any set of geometric vectors containing $\underline{i}, \underline{j}$ and $\underline{k}$ and other geometric vector(s) would also span the space, but it would not form a basis, since such a set would be linearly dependent (the other geometric vector(s) would be redundant).

Exercise 1

Show that the set $\left\{ \begin{pmatrix} 1 \\ 0 \\ 0 \end{pmatrix}, \begin{pmatrix} 1 \\ 1 \\ 1 \end{pmatrix}, \begin{pmatrix} 0 \\ 0 \\ 1 \end{pmatrix} \right\}$ is a basis for the set of all triples of real numbers. ∎

As a result of this last exercise we have two distinct bases, namely

$\left\{ \begin{pmatrix} 1 \\ 0 \\ 0 \end{pmatrix}, \begin{pmatrix} 0 \\ 1 \\ 0 \end{pmatrix}, \begin{pmatrix} 0 \\ 0 \\ 1 \end{pmatrix} \right\}$ and $\left\{ \begin{pmatrix} 1 \\ 0 \\ 0 \end{pmatrix}, \begin{pmatrix} 1 \\ 1 \\ 1 \end{pmatrix}, \begin{pmatrix} 0 \\ 0 \\ 1 \end{pmatrix} \right\}$ for the same vector space, the space of all triples, and in this case both bases consist of three vectors. In fact, although we shall not prove it here, this always happens: for any two sets of base vectors for the same vector space, there are always the same number of vectors in each basis. This enables us to make the following definition.

If $\{v_1, v_2, \ldots, v_n\}$ is a basis for a vector space V, then we say that the vector space is of dimension n.

(*continued on page 47*)

Solution 22.2.2.4

(i) The set of triples is linearly dependent; for instance,

$$\begin{pmatrix} 1 \\ 0 \\ 0 \end{pmatrix} = \begin{pmatrix} 1 \\ -1 \\ 0 \end{pmatrix} + \begin{pmatrix} 0 \\ 1 \\ 0 \end{pmatrix}.$$

(ii) The set of functions is linearly independent, because $\alpha f + \beta g = \underline{0}$ implies that $\alpha x + \beta x^2 = 0$ for all values of x, and this is possible only if $\alpha = \beta = 0$.

(iii) The set of triples is linearly independent.

$$\alpha_1 \begin{pmatrix} 2 \\ 0 \\ 0 \end{pmatrix} + \alpha_2 \begin{pmatrix} 0 \\ 3 \\ 0 \end{pmatrix} + \alpha_3 \begin{pmatrix} 0 \\ 0 \\ 5 \end{pmatrix} = \begin{pmatrix} 0 \\ 0 \\ 0 \end{pmatrix}$$

$$\Rightarrow \qquad \begin{pmatrix} 2\alpha_1 \\ 3\alpha_2 \\ 5\alpha_3 \end{pmatrix} = \begin{pmatrix} 0 \\ 0 \\ 0 \end{pmatrix}$$

$$\Rightarrow \alpha_1 = \alpha_2 = \alpha_3 = 0 \qquad \blacksquare$$

Solution 22.2.2.5

Using the axioms of a vector space, we can show that the given equation is equivalent to

$$(\alpha_1 - \beta_1)\underline{v}_1 + (\alpha_2 - \beta_2)\underline{v}_2 + \cdots + (\alpha_n - \beta_n)\underline{v}_n = \underline{0}.$$

Since the set of vectors $\{\underline{v}_1, \underline{v}_2, \ldots, \underline{v}_n\}$ is linearly independent, the coefficients of the vectors in the above equation are all zero, so

$$\alpha_1 - \beta_1 = \alpha_2 - \beta_2 = \cdots = \alpha_n - \beta_n = 0,$$

which proves the required result. $\qquad \blacksquare$

Solution 22.2.2.6

Suppose, in contradiction to what we want to prove, that the subset $\{\underline{v}_1, \underline{v}_2, \ldots, \underline{v}_k\}$ is linearly dependent; then there are numbers $\alpha_1, \alpha_2, \ldots, \alpha_k$ (not all zero) such that

$$\alpha_1 \underline{v}_1 + \alpha_2 \underline{v}_2 + \cdots + \alpha_k \underline{v}_k = \underline{0}.$$

Therefore

$$\alpha_1 \underline{v}_1 + \alpha_2 \underline{v}_2 + \cdots + \alpha_k \underline{v}_k + 0\underline{v}_{k+1} + \cdots + 0\underline{v}_n = \underline{0}.$$

But not *all* the $\alpha_1, \ldots, \alpha_n$ are zero, and hence the set of vectors $\{\underline{v}_1, \underline{v}_2, \ldots, \underline{v}_n\}$ is linearly dependent — which is a contradiction. $\qquad \blacksquare$

Solution 22.2.2.7

If $\{\underline{v}_1, \underline{v}_2, \ldots, \underline{v}_n\}$ is linearly dependent, then there are numbers $\alpha_1, \alpha_2, \ldots, \alpha_n$, not all zero, such that

$$\alpha_1 \underline{v}_1 + \alpha_2 \underline{v}_2 + \cdots + \alpha_n \underline{v}_n = \underline{0}.$$

Hence

$$\alpha_1 \underline{v}_1 + \alpha_2 \underline{v}_2 + \cdots + \alpha_n \underline{v}_n + 0\underline{w} = \underline{0}.$$

Not all the coefficients in this last equation are zero, and so we have proved the required result. $\qquad \blacksquare$

Solution 1 **Solution 1**

Any triple

$$\begin{pmatrix} x_1 \\ x_2 \\ x_3 \end{pmatrix} = (x_1 - x_2)\begin{pmatrix} 1 \\ 0 \\ 0 \end{pmatrix} + x_2\begin{pmatrix} 1 \\ 1 \\ 1 \end{pmatrix} + (x_3 - x_2)\begin{pmatrix} 0 \\ 0 \\ 1 \end{pmatrix}.$$

Also the set of triples is linearly independent:

$$\alpha_1\begin{pmatrix} 1 \\ 0 \\ 0 \end{pmatrix} + \alpha_2\begin{pmatrix} 1 \\ 1 \\ 1 \end{pmatrix} + \alpha_3\begin{pmatrix} 0 \\ 0 \\ 1 \end{pmatrix} = \begin{pmatrix} 0 \\ 0 \\ 0 \end{pmatrix}$$

$$\Rightarrow \quad \begin{pmatrix} \alpha_1 + \alpha_2 \\ \alpha_2 \\ \alpha_2 + \alpha_3 \end{pmatrix} = \begin{pmatrix} 0 \\ 0 \\ 0 \end{pmatrix}$$

$$\Rightarrow \alpha_1 = \alpha_2 = \alpha_3 = 0.$$

It follows that the given set of triples is a basis. ∎

(continued from page 45)

If it is impossible to find a finite number of elements of a vector space V which form a basis for V, and $V \neq \{0\}$, then we say that V has *infinite dimension*.

It is in fact also true that *any* set of n linearly independent vectors in a vector space V of dimension n is a basis for V, but the proof of this result will also need to be deferred to a later course.

If we assume these results, then we can see that, since $\left\{ \begin{pmatrix} 1 \\ 0 \end{pmatrix}, \begin{pmatrix} 0 \\ 1 \end{pmatrix} \right\}$ is a basis of the vector space of ordered pairs of real numbers, this vector space is therefore of dimension 2. The set $\left\{ \begin{pmatrix} 1 \\ 0 \\ 0 \end{pmatrix}, \begin{pmatrix} 0 \\ 1 \\ 0 \end{pmatrix}, \begin{pmatrix} 0 \\ 0 \\ 1 \end{pmatrix} \right\}$ is a basis for the set of ordered triples, and this vector space is therefore of dimension 3.

Let us look now at some non-geometric examples.

Example 1 **Example 1**

The set of all polynomial functions of degree 2 or less, i.e. of the form:

$$f : x \longmapsto ax^2 + bx + c \qquad (x \in R)$$

where $a, b, c \in R$, forms a vector space with the operations of addition of functions and multiplication of a function by a real number.

We can find many sets of three vectors in this vector space which are linearly independent. One such set, which is particularly simple, consists of the vectors

$$f_1 : x \longmapsto 1 \qquad (x \in R),$$
$$f_2 : x \longmapsto x \qquad (x \in R),$$
$$f_3 : x \longmapsto x^2 \qquad (x \in R).$$

Any other quadratic function can be expressed in terms of these three, and hence they form a basis for the vector space. The dimension of the space is therefore 3. The function

$$f : x \longmapsto 3x^2 - 2x + 4 \qquad (x \in R)$$

can be written as a linear combination of the base elements:

$$\underline{f} = 3\underline{f}_3 - 2\underline{f}_2 + 4\underline{f}_1$$

(We have underlined the f's because we want to emphasize the fact that we are considering the functions to be elements of a vector space.) ■

Example 2

Example 2

We stated in Example 22.2.1.4 that any solution of the equation

$$D^2 f(x) - 3Df(x) + 2f(x) = 0$$

can be expressed in terms of the two solutions

$$\underline{f}_1 : x \longmapsto e^x \qquad (x \in R),$$
$$\underline{f}_2 : x \longmapsto e^{2x} \qquad (x \in R).$$

In other words, these two functions span the space of solutions. Since the two solutions, $\underline{f}_1$ and $\underline{f}_2$ are independent, the set of all solutions of the equation forms a vector space of dimension 2. (If you are worried about the statement that $\underline{f}_1$ and $\underline{f}_2$ are independent, you might like to try to prove it.) ■

22.3 SUMMARY

We have seen how the concept of a vector space may be regarded as having its origins in geometry. Using the concept of a *geometric vector*, we were able to construct an algebraic structure by introducing "addition" and "multiplication by a scalar" on the set of all geometric vectors in two (or three) dimensions. We then saw that we could construct a very similar algebraic structure on the set of ordered pairs (or triples) of real numbers.

There are in fact many different mathematical systems which have the same structure, and we chose to extract the important properties, then to study the abstract structure which possesses these properties. Such a structure we called a *vector space*. The concept of a vector space is of fundamental importance; indeed, so important that we shall build a second level course round it. Clearly, in that course, our first task will be to put the subject on firm foundations and to verify the unproved statements in section 22.2 of this text.

In the next unit on linear algebra we shall discuss what happens when vector spaces are mapped to vector spaces, and one of the important questions will be: "What happens to the dimension of a vector space under a morphism?"

22.4 APPENDIX

Applications of Geometric Vectors

This section is an optional part of the text. It has been included to give a glimpse of the many applications of geometric vectors.

The applied mathematician uses geometric vectors in essentially two distinct ways. First he makes use of them when constructing a mathematical model of a physical quantity such as *force*; and, secondly, he uses them as a convenient means for specifying the position of a point in space.

Mathematical Modelling

The first stage of a problem in applied mathematics is usually one of simplification. We reduce the problem to a reasonable form by making certain assumptions, and then we construct a mathematical model.

This brings us back to the example we discussed in the Introduction (section 22.1.0) of an aeroplane flying in a cross wind. Suppose that the pilot points the nose of the aircraft due north and that his air speed indicator reads 120 mile/h. (This means that, if the air were absolutely still, the aircraft would be moving at 120 mile/h due north relative to the ground.) Let us also suppose that there is a constant easterly wind of 50 mile/h. What is the velocity of the aircraft relative to the ground?

Let us model the velocity of the aircraft relative to the wind by a geometric vector V; then we could represent the physical situation by the following diagram.

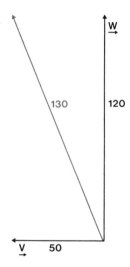

We have our model, but can we draw any conclusions? As the pilot tries to fly his aircraft due north it is constantly carried to the west at a rate of 50 mile/h. The resulting velocity relative to the ground can be modelled by the sum $V + W$, which in this case implies that the aircraft has a speed of 130 mile/h relative to the ground in the direction indicated on the diagram.

There is one extremely important point to notice about this example. We can model the individual velocities by geometric vectors, but in order to draw our final conclusion we need to assume that the addition operation for geometric vectors is the appropriate model of physical combinations of velocities.

Physical Situation	Mathematical Model
Velocity of the wind	$\underset{\rightarrow}{W}$
Velocity of aircraft	$\underset{\rightarrow}{V}$
Combination of velocities	$+$

This, of course, depends on experimental verification or valid deduction from previously validated models. In this case the model apparently works, and so the physical and mathematical situations are related by a morphism.

Forces

Geometric vectors are often used to model *forces*. (We shall assume only that you have an intuitive notion of the meaning of the word *force*.) Suppose, for example, that two tug-boats are towing a ship.

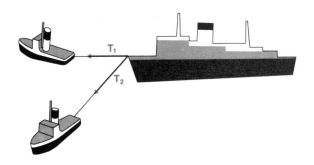

We might make the gross simplification that the ship is a particle, and then represent the two forces in the tow-ropes by geometric vectors. What is the resulting force? The important point is that we can verify by experiment that forces on a particle are combined in the same way as geometric vectors are combined by addition. In other words, geometric vectors are an adequate representation in that their rule of combination corresponds to the physical combination of the quantities which they represent. To find the resulting force we simply *add* the geometric vectors, and this gives us the appropriate model of the net force.

It isn't always appropriate to model forces by geometric vectors. For example, if we wish to take account of the turning effect of the tow-ropes on the ship, then the points at which they are attached will clearly be important. In this case we might simplify the ship not to a particle, but to a line segment, and we might model the forces by arrows attached to the appropriate points on the line segment.

Geometric Applications

The second important application of geometric vectors is their ability to specify points in space relative to a fixed point (called the origin). If we use a geometric vector $\underline{r}$ to define a translation, then the image of the origin $\underline{r}(O)$ is uniquely determined. This is the point P say.

It may be rather difficult at this stage to see the advantages of determining the point P by the geometric vector r rather than by, say, its co-ordinates (x, y, z) in a Cartesian co-ordinate system. (See figure below.)

Often we wish to determine the position of a point in space in a problem which has arisen from a physical situation, and in which we have used geometric vectors to model quantities such as force. There are definite advantages in keeping all our discussion in terms of geometric vectors in this case. We could manage without geometric vectors, just as we could manage without (school) algebra and use arithmetic only. But there are considerable advantages to be gained in the statement of problems and in the manipulation required to solve them, if we use geometric vectors throughout.

As a very trivial example of the simplification which the use of geometric vectors can bring to a problem, consider the formula for the mid-point M of the line joining points P and Q.

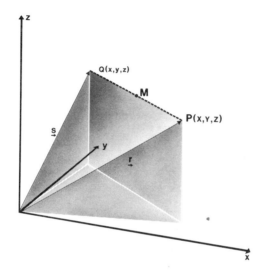

The Cartesian co-ordinates of M are

$$\left(\frac{x + X}{2}, \frac{y + Y}{2}, \frac{z + Z}{2}\right),$$

whereas the point is equally well determined by the geometric vector $\frac{1}{2}(r + s)$. We require a third of the time and space to convey exactly the *same* information if we use geometric vectors.

Think for a few moments how you would convey the information on the following diagram (in which r determines the point P, and $\vec{F}$ is used to model a force applied to a particle at P) in terms of Cartesian co-ordinates only.

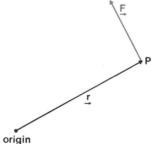

Postscript

"The human mind has never invented a labor-saving machine equal to algebra."

The Nation, Vol. 33, p. 237

Unit No.		Title of Text
1		Functions
2		Errors and Accuracy
3		Operations and Morphisms
4		Finite Differences
5	NO TEXT	
6		Inequalities
7		Sequences and Limits I
8		Computing I
9		Integration I
10	NO TEXT	
11		Logic I — Boolean Algebra
12		Differentiation I
13		Integration II
14		Sequences and Limits II
15		Differentiation II
16		Probability and Statistics I
17		Logic II — Proof
18		Probability and Statistics II
19		Relations
20		Computing II
21		Probability and Statistics III
22		Linear Algebra I
23		Linear Algebra II
24		Differential Equations I
25	NO TEXT	
26		Linear Algebra III
27		Complex Numbers I
28		Linear Algebra IV
29		Complex Numbers II
30		Groups I
31		Differential Equations II
32	NO TEXT	
33		Groups II
34		Number Systems
35		Topology
36		Mathematical Structures